Homework Helpers: Biology

Biology

Matthew Distefano

Career Press
Pompton Plains, NJ

HOMEWORK HELPERS: BIOLOGY
TYPESET BY EILEEN MUNSON
Original cover design by Lucia Rossman, Digi Dog Design
Printed in the U.S.A.

To order this title, please call toll-free 1-800-CAREER-1 (NJ and Canada: 201-848-0310) to order using VISA or MasterCard, or for further information on books from Career Press.

The Career Press
220 West Parkway, Unit 12
Pompton Plains, NJ 07444
www.careerpress.com

Library of Congress Cataloging-in-Publication Data

Distefano, Matthew, 1967-
Homework helpers. Biology / by Matthew Distefano. -- Rev. ed.
p. cm.
Includes bibliographical references and index.
ISBN 978-1-60163-164-0 (alk. paper) -- ISBN 978-1-60163-662-1 (ebook : alk. paper) 1. Biology. I. Title.

QH307.2.D57 2011
570--dc23

2011036955

Dedication

I wish to dedicate this biology review book to my 8th grade science teacher, Mr. Comer, who was the first to instill in me the love and wonder of science. Mr. Comer showed me that proper planning and taking the right approach could make all the difference.

~

Acknowledgments

To Greg Curran for giving me the opportunity to be part of this book project and for the many insights he has provided me over my years of teaching.

To Jessica Faust whose patience and drive has made the idea for this book become a reality.

To my wife, Shawn, and my children, Amanda and Luke, for all the time and support they provided to me towards the completion of this book.

To my mom for all the help with Amanda and Luke she provided during the creation of this book.

To Mike Curtain and Jeff Butkowski for providing all the computer and tech support needed on this project.

To Laura Prevatali and John Distefano for all the artwork featured in this project.

To the faculty members of the Fordham Preparatory School, who have been role models of good teaching, for their constant display of the passion needed to be successful in this profession.

CONTENTS

CONTENTS

Welcome to *Homework Helpers: Biology*!

Homework Helpers: Biology was written with the student in mind. I hope to make the study of biology easier. This book provides clear, concise explanations of all the major topics presented in a biology course. With this book, we hope to simulate the feel of one-on-one tutoring sessions with a teacher. Many of the tables throughout the book have been created to simulate the blackboard in a classroom, recreating a familiar learning environment. Numerous practice questions have been provided, and all questions in this book come with answers and explanations.

When faced with taking a biology course, many students can easily become intimidated. Biology is a subject full of scientific terms and jargon that can make understanding and studying the concepts of the course very difficult. In this book, I hope to make these terms and concepts easier to understand through the use of diagrams, examples, and easy-to-understand text. Although memorization does play a role in a course such as biology, comprehension of the material leads to much longer recall-ability and much more enjoyment of the subject matter. Do not read this review book from cover to cover. After each topic is presented to you in your biology class, review and enforce the material by completing the corresponding sections from this book. Focus on the boldfaced words in each section and be sure to know their definitions and connection to the chapter topic. Continually test yourself with the questions found at the end of each review section. If you answer a question incorrectly, go back and review that particular area. All questions in this book come with answers that have complete explanations. Even incorrect answers for each question are discussed so that you will know why certain responses are not correct. Be sure to not only review correct answers, but incorrect responses as well. As much can be learned from the knowledge obtained by making an incorrect choice as from the knowledge obtained from making a correct choice. Remember, good comprehension of a subject matter only comes after sufficient review. Work on the material from this review book for a short period of time every night. This is always the best way to approach a course with as much to learn as biology. Make ***Homework Helpers: Biology*** the book that helps brings comprehension, confidence, and enthusiasm to your study of biology.

Confidence and enthusiasm about a subject comes through review and practice, and this book provides a medium for this to happen. I hope you enjoy this book, and more importantly, the subject of biology.

1

Studying Biology

In this opening chapter, you will be introduced to the subject of biology, and you will learn how to think like a scientist and about the many tools scientists use. Biology is the study of life, and so this chapter concludes with some discussion about what life is and how living things are identified.

Lesson 1-1: Thinking Like a Scientist

In order to study biology, one must learn something of the skills of the biologist. It just happens that one of the most important skills that scientists have is something that most young children practice on a regular basis: the act of questioning. Progress in the field of biology depends on people who ask questions about the living world around them and take time to test these questions. Why is human blood red? Why do elephants have long ivory tusks? Why are most plants green? To answer questions about life, many approaches are possible. But there are steps that are common to almost all of theses approaches. These common procedures that biologists use to gather information in order to answer questions is called the **scientific method**. The scientific method can be looked at as an organized way of solving a problem. When using the scientific method you will follow certain steps in order to try to solve or explain unknowns in biology.

Steps of the Scientific Method

1. Recognizing a problem or something that needs to be solved.
2. Researching anything that may be known about the problem.
3. Proposing a possible answer, sometimes called a hypothesis, that could be a solution to the problem. A hypothesis must be testable by experimentation to be valid.
4. Conducting an experiment to test the proposed hypothesis to the problem.
5. Collecting and analyzing data from the experiment that will help to determine if the hypothesis is correct or incorrect.
6. Verifying if the hypothesis you stated is supported or refuted by the data from the experiment.
7. Making sure the experiment you conducted can be done again with the same results.
8. Sharing the results of the experiment with others.

••••••

The following is an example of a problem solved using the steps of the scientific method:

Example

1. **Problem or unknown.**

   A gardener buys two fertilizers for his tomato plants (fertilizer A and fertilizer B). This gardener wants to know which fertilizer will make his tomato plants produce more fruit.

2. **Researching what is known about the problem.**

   The gardener reads the labels of each fertilizer and tries to find any information he can on the Internet about these two brands of fertilizer.

3. **Proposing a possible answer (hypothesis).**

   The gardener decides that fertilizer A will probably work better based on the research he has done on the two products.

4. **Conducting an experiment for a possible solution to the problem.**

   The gardener uses fertilizer A on five of his tomato plants and fertilizer B and the other five tomato plants he has in the garden. He also grows five plants with no fertilizer at all. The gardener is careful to treat the 15 tomato plants in the same manner (watering, exposure to sunlight, soil), except for the use of the different fertilizers in the first two groups. The fertilizer used in each part of the experiment would be called the **variable** because it is the part of the experiment that is being questioned or studied. The five tomato plants not being given the fertilizer act as a **control group**. The control group acts as the basis of comparison for the variable being tested in the experiment.

5. **Collecting and analyzing data from the experiment.**
   After a two-week period, the gardener checks the amount of fruit on the 15 plants in the study. Because the gardener is counting the number of fruit on each plant, this is considered **quantitative data**. Quantitative data is information that involves numbers. Data that would not involve numbers, but instead involves such things as a description, is called **qualitative data**. The results show that the five plants grown with fertilizer B have almost twice as much fruit as the five plants grown with fertilizer A. The 5 plants grown without any fertilizer have the least amount of fruit.

6. **Drawing conclusions based on the data collected and analyzed.**
   The gardener saw that his initial hypothesis that fertilizer A would work better to produce more fruit than fertilizer B was wrong. Even though his hypothesis was wrong, the gardener learned valuable information that he can now apply to the success of his garden.

7. **Making sure the experiment and results are repeatable.**
   The gardener tested these two fertilizers on five plants each within the garden, and all showed that fertilizer B produced more fruit than fertilizer A.

8. **Sharing the results with others.**
   The gardener showed all his friends who grew tomatoes the wonderful results he had obtained in his garden with the use of fertilizer B.

••••••

As demonstrated in the example, the scientific method is an organized approach to solving a problem. Most people do it all the time in everyday situations of life, without even being aware they are performing these procedures. I will give another example to prove my point. When you cross a busy street you are unconsciously performing the steps of the scientific method. When you arrive at the curb of the street that you must cross, you are faced with the problem: I must get across this street without getting hit by a car. Your brain quickly recalls other times you have crossed a street and the steps you took to accomplish this task. You then propose a possible answer to this problem, namely how quickly you will have to walk and at what precise time, in order to reach the other curb safely. In order to propose a hypothesis about how to cross the street safely, you would collect information about the street and where the cars were coming from—from what direction and their speeds. The experiment would involve you actually walking across the street. Your hypothesis about how to cross the street safely would be proven correct if you got to the other side of the street safely. If you did not get across the street safely, your hypothesis would be incorrect. And just like in any biology experiment, an experiment that proves a hypothesis wrong can be used to collect as much information about the problem as an experiment that proves the hypothesis correct. Though, in this particular case, if your educated guess about how to cross the street is wrong, you may be considering a new hypothesis in the hospital!

It is important to understand that doing an experiment cannot always test a particular question that is asked about something in biology. Some problems can only be observed and possibly solved by doing what is called ***fieldwork***. Fieldwork is when a scientist observes an organism in the wild and collects data about their observations. For example, suppose a scientist wants to know where gray squirrels prefer to hide acorns in the forest. Setting up an experiment in a lab would be almost impossible for this particular question. The best way to approach this question would be to observe the squirrel in the natural environment (fieldwork), and collect data concerning the food storage habits witnessed.

Scientific Theory

To best understand the meaning of a scientific theory it is important to first understand that there is a major difference between a scientific theory and the word "theory" as it is many times used in a nonscientific manner. The word "theory" is used regularly to express a "guess" someone may have concerning a particular situation. For example, in the mall you may overhear one teenage boy tell his friend his "theory" about why their dates did not show up to meet them as expected for lunch. In science, a theory has a much different meaning. A **scientific theory** is an explanation for something that has occurred in nature that has been substantiated by a large amount of data that has been collected from many different experiments from many different experimenters. A scientific theory is not an absolute truth but is considered to be the best possible explanation to a certain question available at that time based on the experiments and data collected. As new data is collected and new experiments are conducted, a current theory can be altered, revised, completely abandoned, or left alone. For example, many scientists believe that a huge asteroid struck the earth about 65 million years ago and hit what is now the Yucatan Peninsula in Mexico, and that this event triggered the extinction of the dinosaurs. With the evidence that is present at this time concerning this issue, the asteroid is a theory about how the extinction of these giant reptiles occurred.

The International System of Units

As seen in the previous examples, many times the data collected by a scientist involves numbers and is called quantitative data. Scientific data involving numbers usually has units associated with the number. A unit tells you what type of quantity the number given represents. Units in science follow a system agreed upon by the countries of the world and by all scientists called the International System of Units. The International System is a decimal system of measurement whose units are based on scaled multiples of 10. The math involved with using this system is very easy because by simply multiplying or dividing by bases of 10, you can determine all your final answers. Some of the most commonly used System International (SI) units are in Table 1.1.

Commonly Used SI Units

Length	**Temperature**
1 meter = 1,000 millimeters	0° Celsius = freezing point of water
1 meter = 100 centimeters	100° Celsius = boiling point of water
1,000 meters = 1 kilometer	Formula to convert Fahrenheit to Celsius: C = 5/9(F – 32)
Volume	**Mass**
1 liter = 1,000 milliliters	1 gram = 1,000 milligrams
1 milliliter = 1 cubic centimeter	1,000 grams = 1 kilogram

Table 1.1

Sometimes, in biology class, you will have to convert from one of these given units to another. For example, in order for a student to determine the length of an object seen under a compound microscope, the length of the diameter of the ***field of view*** must be measured. Field of view is the visible area that can be observed through a microscope lens. This measurement is usually done with a millimeter ruler. Millimeters is a good unit for measuring the length of the field of view, but is usually too large to measure the length of organisms or objects seen within the field of view. So, many times, the length of the field of view is converted into a unit called ***micrometers***. Using the conversion factors from the table of commonly used SI units, the conversion could be done in the manner shown in the following example.

Example

A student measures the diameter of the field of view of his compound microscope using a millimeter ruler and determines the length to be 3.5 millimeters.

Later in the biology lab class, the same student, using the same microscope, notices a small microorganism moving across the field of view.

The student is then asked to determine the length of the microorganism in micrometers.

Answer:

Because the microorganism takes up about one-quarter of the distance across the diameter of the field of view, the organisms must have a length of about 1/4 (3.5 millimeters), or about 0.875 millimeters.

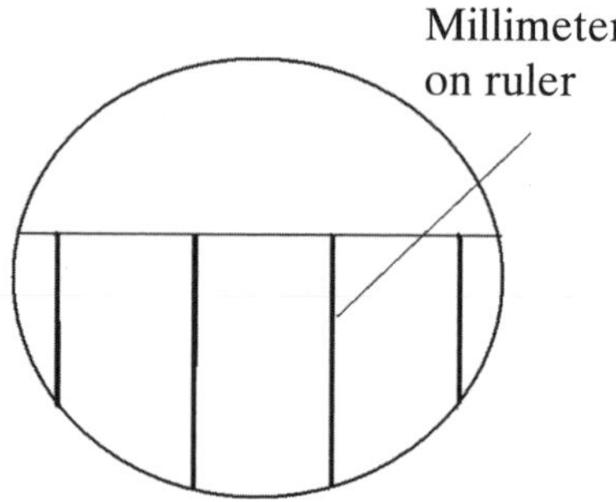

Field of View

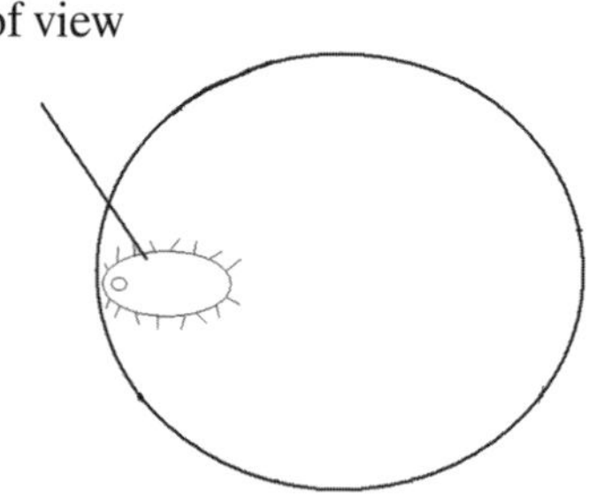

Field of View

Because we know that 1000 micrometers is equal to 1 millimeter, to convert this measurement into micrometers you would multiply 0.875 by 1000 and obtain the number 875 micrometers. This same answer could also be converted to the proper units using a mathematical proportion.

$$\frac{1 \text{ millimeter}}{1{,}000 \text{ micrometers}} = \frac{0.875 \text{ millimeters}}{x \text{ micrometers}}$$

If you cross multiply in the math proportion above and solve for the unknown (x), you will obtain the same answer as before: 875 micrometers.

••••••

Many times it is easier to use the standard abbreviations for these units, instead of spelling out the entire word each time.

Abbreviations of the Most Commonly Used International System Units

Unit	Abbreviation	Unit	Abbreviation
gram	g	centimeter	cm
meter	m	kilogram	kg
liter	l	milligram	mg
millimeter	mm	milliliter	ml
kilometer	km	cubic centimeter	cm^3

Table 1.2

Practice Section 1-1

1. Using what you have learned and reviewed about the scientific method from this section of the book, write out step-by-step scientific method solutions to each of these problems. Be sure that each step in your solution is connected to the previous step in the scientific method.
   a. Do colorful male birds more successfully attract female birds for mating than less colorful male birds?
   b. Do carpenter ants prefer to make their nests in oak or pine wood?
   c. Do rose bushes grow better in acidic soil or alkaline soil?
2. Using the unit conversion chart shown in Table 1.1, complete some of the following conversions:

4,000 millimeters	=	________	meters
35.6 kilometers	=	________	meters
77° Fahrenheit	=	________	Celsius
287 milliliters	=	________	liters
599 grams	=	________	kilograms
4 liters	=	________	cubic centimeters

3. What is a scientific *hypothesis*? How is a hypothesis different from a *theory*?
4. Give a possible hypothesis for the following problem: You walk into the living room of your house and click the remote button to turn on your TV. The TV does not turn on."

Lesson 1-2: Science Equipment

In order to properly study life on this planet, some basic equipment is needed. Most science equipment is used to aid the human senses in the many observations that must be done during experimentation. Modern science equipment is used to enhance the human senses, such as the ability of the human eyes to see and the human ears to hear.

Microscopes

Probably one of the most important and most used pieces of equipment of the biologist, or the high school or college science student is the microscope. The microscope better enables scientists to study the microscopic world of Earth, where many organisms dwell and the structures of life reside. The major goal of the microscope is to magnify organisms that are too small for the human eye to see clearly. There are many different kinds of microscopes. The kind of microscope you probably have in your lab class is a **compound microscope**. This type of microscope is called ***compound*** because the microscope focuses and magnifies an object with two lenses. The power of the first lens, or ***ocular***, that you look through is multiplied by the power of the second lens, or ***objective***, to obtain the total magnification of the object being studied. Many compound microscopes usually have two or more objective lenses, but only one of these lenses is used at any one time in conjunction with the eyepiece lens.

Example

A compound microscope has an eyepiece (ocular) of 10X and two objective lenses of 10X and 40X. What would be the total magnification of an object viewed through these two lenses?

Answer:

There are two possible answers to this question based on which objective lens is used.

If the microscope is adjusted to use the 10X lens:
Eyepiece lens (10X) × objective lens (10X) = total magnification (100X)

or

Eyepiece lens (10X) × objective lens (40X) = total magnification (400X).

••••••

Many structures and organisms can only be studied properly under a microscope because they are too small for the human eye to see. Compound microscopes usually have total magnification powers ranging from 100X to about 2000X depending on the quality of the instrument.

The image is focused by use of the **coarse focus** and **fine focus** knobs on the microscope. The coarse focus knob makes large adjustments to the focus and the fine focus knob makes small adjustments to focus.

The **diaphragm** on the microscope adjusts the amount of light that will pass into the lenses for viewing.

Compound Microscope

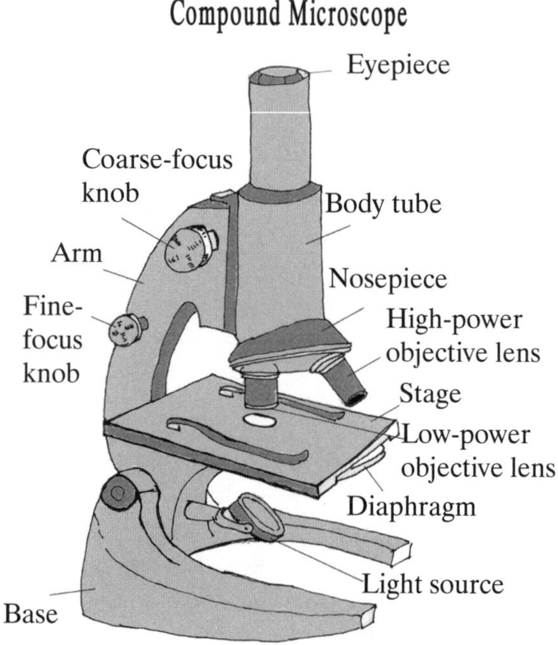

Another characteristic of compound microscopes is resolving power. **Resolving power** is the ability to distinguish between two points that are very close together. Simply stated, resolving power means how *clear* a microscope makes the object you are trying to magnify and view. The better the quality of the microscope, the smaller the object will be that can still be brought into a clear view.

There are some objects so small that even the compound microscope cannot bring them into proper magnification and resolution. There is another type of microscope that can help in this regard called the **electron microscope**. The electron microscope has the ability to magnify an object up to 500,000 times its actual size. This microscope does not use light like the compound microscope, but instead uses a beam of electrons for magnification. Some shortfalls with the electron microscope include high cost and limitations on what can be viewed with this instrument. With an electron microscope, air affects the movement of electrons, which affects the magnification. This means that all objects must be placed in a vacuum before they are viewed. Consequently, only inorganic or dead objects can be magnified with an electron microscope.

Balances

Many times during lab experiments in biology, objects will have to be massed. The use of various types of **balances** has greatly increased the accuracy of these measurements in biology. The **triple-beam balance** is very common in high school lab classes. The balance works by moving what are called ***riders*** across the upper beams of the balance until the pointer of the balance reads level. At this point the numbers under the placement of the riders will read the mass of the object sitting on the pan of the balance. The following diagram gives an illustration of how this would be done.

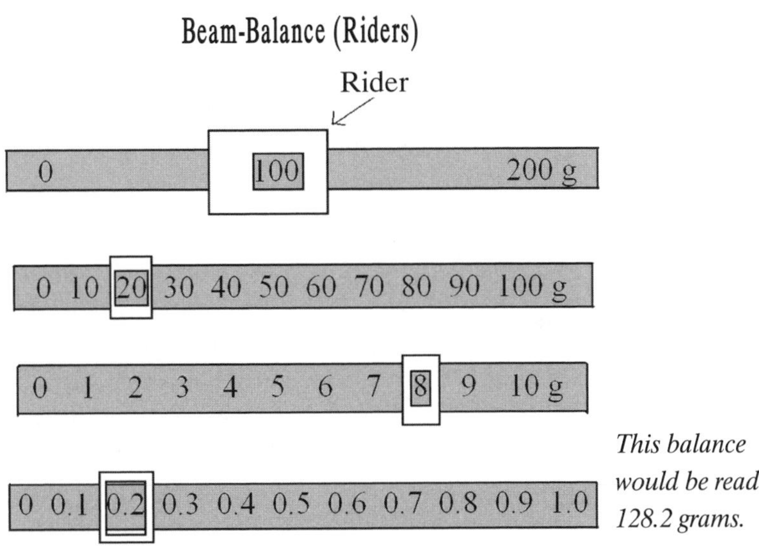

This balance would be read 128.2 grams.

Electronic balances are a type of balance that enables scientists to obtain a very accurate mass measurement with much less effort then the triple-beam balance. All that has to be done with an electronic balance is turn it on, wait until it reads zero, and then place the object to be massed on the pan.

Electronic Balance

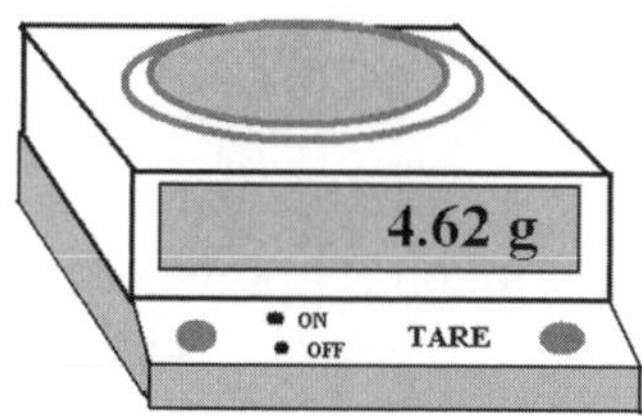

A precaution to take with all balances is to never place the object to be massed directly onto the pan. Protect the pan of the balance by placing the object to be massed in a ***weighing boat*** or ***dish***. When using a triple-beam balance, you must mass the weighing boat first and then subtract this amount from the mass of the object and the weighing boat to obtain the final mass of the object. When using an electronic balance, place the weighing boat on the pan and hit the ***tare*** button. The ***tare*** button will re-zero the balance. You can now fill the weighing boat with the substance you need to mass and the electronic balance will read only the mass of this substance.

Tools for Separating Chemical Mixtures

An **ultracentrifuge** is an instrument used to separate substances in a solution by density. Usually, test tubes full of organic tissues are placed in an ultracentrifuge and then spun at very high speeds. This spinning causes the tissue sample to separate in the tube based on weight with the heaviest or densest substances ending up at the bottom of the tube and the least dense at the top. One particular use of the ultracentrifuge is to separate the different substances found inside a cell into various zones within the test tube based on the weight of these individual parts.

Centrifuge Test Tube

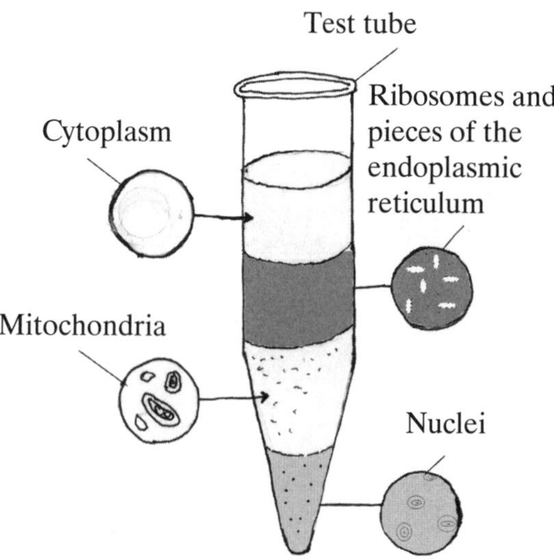

Another way to separate molecules by size is through a tool called ***gel electrophoresis***. A gel is prepared that has a series of narrow grooves along the surface. The chemical being tested is specially treated to give it a negative charge. The treated chemical is then poured into the grooves that run in straight lines across the surface of the gel. The gel is then placed between two electrodes and electricity is run through the gel, which causes one side of the gel to become positive and the other side of the gel to become negative.

After the electrophoresis equipment is turned on, the positively charged chemicals in the grooves move toward the negative side of the gel and negatively charged chemicals move toward the positive side of the gel. The smallest particles in the substance being tested move the fastest in the grooves. At the end of the process you should have a separation of molecules by size and charge.

When the chemical being studied with electrophoresis is DNA, special enzymes must first cut the DNA into small pieces. The DNA sample is then placed into grooves along the surface of hardened gel, similar to the gel for the electrophoresis test, and an electric current is run through the gel. This current separates the smallest negatively charged pieces of DNA from the largest negatively charged pieces of DNA along the grooves of the gel. When the process is finished, the gel has a certain unique pattern. This pattern can be studied and compared to other samples of DNA for analysis.

A third way that chemicals are sometimes separated is through a process called **chromatography**. In this process, the chemical being tested, which is usually chlorophyll in biology lab courses, is placed as a liquid dot near the end of a long strip of paper. This paper is then placed in a solvent, usually some type of alcohol, with the chemical dot right above the surface of the solvent. The solvent will soak into the paper and move its way up through the chemical dot. The molecules located inside the chemical dot will be brought up the paper with the chemical solvent. At the end of the process, this long strip of paper will have bands of colors along its surface that show where the solvent deposited certain molecules from the test chemical. The rate at which the molecules are deposited along the strip of paper becomes important for identification of these test chemicals.

Chromatography Equipment

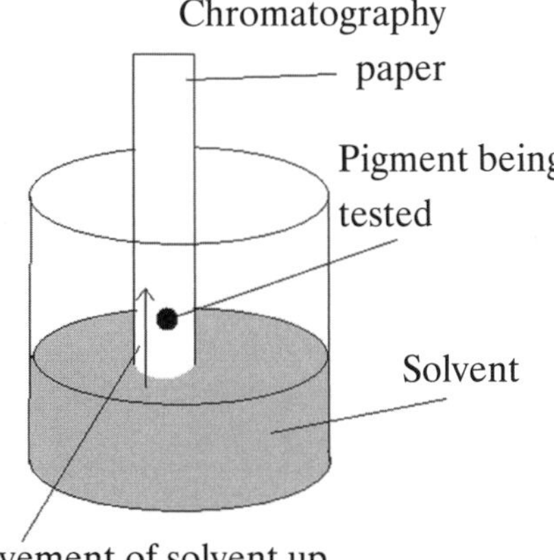

Dissecting Tools

The study of life sometimes involves the dissection of preserved organisms to better understand the appearance and function of their various body parts and how these parts are similar or different from other organisms studied. Some of the most important tools in any dissection are presented in Table 1.3:

Dissection Equipment

Equipment	Function
Scalpel	A sharp blade used to cut open the specimen being studied
Dissecting pan	A low flat pan that is filled with hard wax that provides the work area for the dissection
Dissecting pins	Large T-shaped pins used to secure parts of the specimen to the hard wax surface of the dissecting pan
Tweezers	Used to lift out small parts of the specimen and to help move parts within the body of the specimen for study
Probe or dissecting needle	A thin, metal needle-like structure used to point at hard-to-reach sections of the specimen. The probe is also used to gently remove or "tease" organic tissue from the specimen
Scissors	Used to cut open and remove parts of the specimen for study

Table 1.3

Chemical Indicators

There are many chemicals that will aid you in the study of living organisms in the biology lab class. I have listed some of the most common ones with their most common function. Many of these chemicals are ***indicators***. Indicators are chemicals that change color under specific chemical situations, such as pH level.

Indicators

Chemical	Function
Lugol's iodine	Used to test for the presence of starch
Benedict's solution	Used to test for the presence of simple sugars
Ninhydrin	Used to test for the presence of proteins
Sudan III	Used to test for the presence of lipids
Methylene blue	Used as a biological stain to increase the contrast of some cell parts viewed under the microscope
Phenolphthalein	An indicator that turns from colorless at pH 8.3 to red at pH 10
Bromthymol blue	An indicator that turns from yellow at pH 6 to blue at pH 7.6

Table 1.4

Collecting and Analyzing Data

In the study of biology, many times experiments are conducted or field research is done where large amounts of quantitative data are collected. Usually, the best way to summarize, analyze, and express this data is in charts and graphs.

Many reports and articles presented by biologists concerning issues in life science contain charts and graphs that represent the data that they collected. This data is usually collected to support the hypothesis of the experiment.

Time	Temperature (°C)	Time	Temperature (°C)
6:00 a.m.	13	6:00 p.m.	20
7:00 a.m.	14	7:00 p.m.	20
8:00 a.m.	14	8:00 p.m.	19
9:00 a.m.	15	9:00 p.m.	18
10:00 a.m.	17	10:00 p.m.	16
11:00 a.m.	17	11:00 p.m.	16
12 noon	18	12 midnight	15
1:00 p.m.	19	1:00 a.m.	15
2:00 p.m.	20	2:00 a.m.	14
3:00 p.m.	21	3:00 a.m.	14
4:00 p.m.	21	4:00 a.m.	14
5:00 p.m.	21	5:00 a.m.	13

Table 1.5

It would be much easier to find certain numbers in the table or chart form. It is also easier to see certain patterns that may emerge from the data collected in a particular experiment. It becomes even easier to see patterns that emerge from the same collected data when the numbers are arranged in a graph. See Table 1.6.

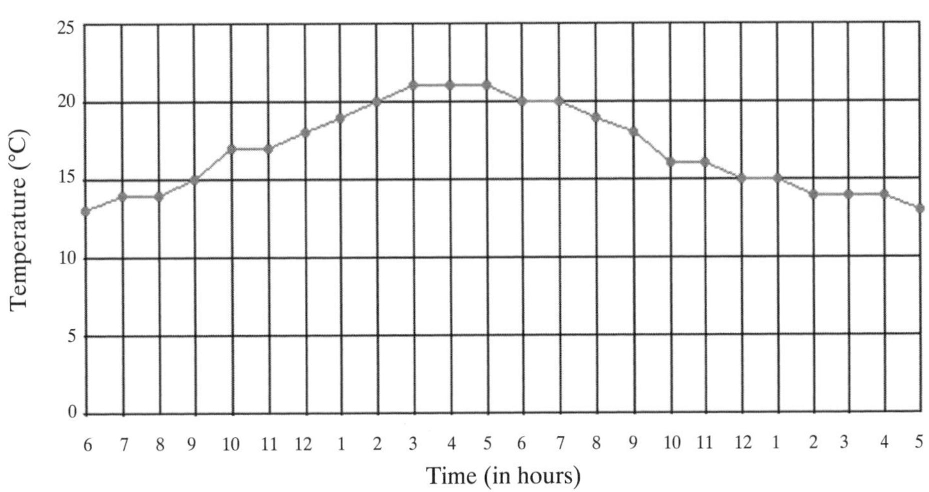

Table 1.6

There are many experiments conducted where the amount of data collected is enormous. For very large amounts of data the advent and use of computers in the lab room has made sorting, analyzing, and collecting data much quicker, more efficient, and accurate. For instance, the human genome project, an experiment where all the genes of a human were mapped, was made possible because of the use of super computers. Super computers were able to deal with the enormous amounts of data much more quickly than if they were not used in such a project.

Practice Section 1-2

1. Read each short story and then record what group of instruments from the following lists would be the most useful in helping to conduct that activity.

Group A	**Group B**	**Group C**
ninhydrin	Benedict's solution	microscope glass slide
test tube	test tube	microscope
hot plate	hot plate	methylene blue
Group D	**Group E**	**Group F**
Lugol's iodine	dissecting pan	meter stick
test tube	scalpel	electronic balance
	probe	weighing boat
	dissecting pins	

Story #1

Peter will be testing the presence of proteins in an unknown substance tomorrow in biology lab class. Which group of instruments would be the most useful in collecting this data?

Story #2

Mary needs to remove the small intestines of a preserved earthworm. Which group of tools would be the most helpful?

Story #3

Billy was asked to mass a block of wood and to calculate its volume. Which group of instruments would help him in this task?

Story #4

Sarah was asked to observe the cell wall of an onion plant cell. Which group of instruments would best help her achieve this task?

2. What makes a microscope a "compound microscope"?

Lesson 1-3: Identifying Life

Biology is the study of life. In order to study life, it is important that we can identify life in the environment. The problem in trying to discern things in the environment as living or not living is not always easy. Living things can, at times, resemble nonliving things, such as a leafless oak tree in the wintertime. Nonliving things can, at times, resemble living things, such as the flame of a match as it uses energy and grows in size to burn. I will attempt to end this confusion by showing you two ways to identify what we call life. Identification of living things can be accomplished by first identifying the actions common to all living organisms, and second by identifying the building block common to the structure of all living organisms.

The Actions of Life

The best way to identify if something is living is by first observing what it does. All living things perform what biologists call life processes. Some nonliving things may appear to perform some of these actions, but only living organisms truly perform them all. There are seven important life processes that seem to best define the actions of life. They are:

7 Life Processes

Nutrition **Transport** **Respiration** **Excretion**
Regulation **Growth** **Synthesis**

••••••

Nutrition is the activities that an organism performs to obtain and use nutrients from the environment. These nutrients are used for energy, growth, and repair of body cells. A squirrel looking for acorns and eating them is an example of this process.

Synthesis is the life process in cells by which small molecules are connected chemically to create larger and more complex chemicals needed by the organism. As a result of synthesis your body is able to change the food you ingest into more of you.

Growth refers to the increase in cell size and the increase in cell number within an organism. If you grew 2 inches taller over the summer, the number of cells has dramatically increased in your body to accomplish this task.

Regulation includes all the processes within an organism that control and coordinate the many activities that a living organism must perform to adjust to the internal and external environment. This is a fancy way of saying that living things respond to *stimuli*. An example of this would be the ability of your body to let you know when you are hungry or thirsty, or when your eyes squint in bright light.

Respiration is the combination of two chemical processes: the removal of energy from food molecules and the conversion of that energy into a more usable form called ATP. When food enters living things, energy must be removed from it in just the right amounts.

Excretion is the removal of waste produced from cellular respiration when food is processed. When food is broken up to release energy and to obtain chemicals for the body, there are other chemicals released during this process not needed by the body. Also, during some chemical reactions in the body, there is an excess of the chemical products created that also must be removed. Excretion rids the living organism of these chemicals before they can poison the organism and upset the internal environment.

Transport is the movement of materials into a cell and within a cell, and the movement of materials throughout an organism. Oxygen entering the cells of your body after you breathe is an example of this process.

Critical to life is its continuance into the future. The continuation of life is made possible by the process of **reproduction**. The reason this process is not grouped with the others is that an individual organism does not have to conduct the process of reproduction to survive, but for life to continue after the individual organism dies, replication of that organism is necessary. Reproduction enables this replication to occur, by the use of a very remarkable chemical called ***DNA***, which we will learn more about in later chapters.

These life processes, or life functions, described here are essential for an organism to be able to adapt to changes in the environment and for the organism to perform the necessary reactions needed to produce organisms for future generations. These actions are found in all living things, from the smallest bacteria to the largest tree. **Metabolism** is all the chemical reactions of life working together to enable an organism to remain alive and, thus, distinguish it as living. It is important that a living thing keeps these chemical processes occurring at the correct rate and time. **Homeostasis** is when living things are able to regulate these chemical processes precisely and maintain a stable internal environment.

Now that we have described the common processes of all living things, we can further identify life by seeing what is common about the make-up of these things.

The Structure of Life

The basic unit of everything that is alive is the **cell**. What this means is that cells are the structure of life. All living things on Earth are made up of one or many trillions of cells, but all are made of cells. To closely examine something in the environment and to recognize it is created of cells informs us that it is, or was, alive. Cells perform the actions of life that were described earlier. In fact, another way to look at these chemical processes is to say that these are actions that all cells perform to stay alive. Living things must perform these processes because all cells must perform these processes. This means that all living things are a function of cells. In Chapter 2 we will look at cells more closely and see how structures within cells are able to perform these many tasks of life.

Practice Section 1-3

1. Read statements *a* through *f* and link each to an important chemical process of life from the word bank.

nutrition	reproduction
transport	synthesis
excretion	growth
regulation	respiration

   a) An earthworm burrows into the cool, moist, underground from the dry, warm, surface.
   b) A small caterpillar feeds on the leaves of an oak tree in the middle of a forest.
   c) After finishing lunch, a high school student's blood distributes food molecules to the many cells inside his body.
   d) A person in a theatre screams after experiencing a scary scene from the movie.
   e) An enzyme is created from smaller chemicals inside the human body.
   f) The puppy that Sarah received for her birthday last year is twice its weight.
2. What are life processes? Can you name them all?
3. What is metabolism? How is metabolism related to the concept of homeostasis?
4. Why is the reproduction not considered one of the life processes?

Terms From Chapter 1

cell	growth	resolving power
chromatography	homeostasis	respiration
compound microscope	hypothesis	scientific method
control group	indicator	synthesis
electron microscope	nutrition	transport
electronic balance	qualitative data	triple-beam balance
electrophoresis	quantitative data	ultracentrifuge
excretion	regulation	variable
	reproduction	

Chapter 1: Exam

Matching Column for Life Processes

Match the life process with its definition.

Life Processes	Definition
1. nutrition	a. The removal of metabolic waste from an organism.
2. synthesis	b. The chemical process of oxidizing organic molecules to release energy.
3. growth	c. The process of obtaining food.
4. transport	d. Combining small molecules to create larger more complex molecules.
5. regulation	e. The increase in cell size and/or number.
6. reproduction	f. The movement of materials within the cell or within the organism.
7. respiration	g. The control and coordination of chemical processes within the organism.
8. excretion	h. The replication of an organism.

Matching Column for Scientific Tools

Match the scientific tools with its function.

Scientific Tool	Function
9. compound microscope	a. An instrument used to separate parts of a cell based on density.
10. chromatography	b. A method of separating small charged chemicals in a gel.
11. scalpel	c. A method of separating chemicals based on the rate they move and saturate a special paper.
12. electrophoresis	d. A device that used two lenses to magnify small images for study.
13. ultracentrifuge	e. A sharp object used to make an incision during a dissection.
14. balance	f. A substance whose color change helps to identify properties of a specific chemical.
15. indicator	g. An instrument used to measure the mass of an object.

Multiple Choice

16. Which of the following would be considered quantitative data?
    a) The color of petals of a flower.
    b) The smell of petals of a flower.
    c) The number of petals on a flower.
    d) All of the above.
17. Which of the following statements would make a good hypothesis?
    a) Is light needed for photosynthesis to occur?
    b) Bacteria are found in the soil layer of undiscovered planets.
    c) Red flowers smell the best.
    d) Sugar levels in human blood increase after eating.
18. 2,000 micrometers are equal to how many millimeters?
    a) 2,000 b) 200 c) 2 d) 0.2
19. Which instrument will be the most helpful in viewing a virus?
    a) compound microscope
    b) electron microscope
    c) ultracentrifuge
    d) electronic balance
20. If a scientist wanted to test the amount of sugar found in a sample of human urine, which of the following indicators would be the most helpful?
    a) ninhydrin c) Benedict's solution
    b) Lugol's iodine d) methylene blue
21. Some amino acid molecules are joined to form large protein molecules. What life process does this illustrate?
    a) synthesis b) transport c) excretion d) respiration
22. A turtle climbs out of cool pond water onto a log to warm itself in the sun. What life process does this illustrate?
    a) growth b) regulation c) respiration d) nutrition

Short Response

23. A doctor gathers a group of 200 people in his study. To begin the experiment he weighs all 200 people. The doctor gives 100 people in this group a newly designed pill to test its effectiveness on weight loss. The doctor does not give this pill to the other 100 people. At the end of one month he weighs all 200 people once more.
    a) What is the significance of not giving 100 people in this study the pill?

24. A lima bean seed is planted in 20 grams of soil, given 4 milliliters (ml) of water and placed in a plant greenhouse at 26° C. Another lima bean seed is planted in 20 grams of soil, given 4 ml of water and placed in another greenhouse at 20° C.
    a) What is the variable in this experiment? How do you identify it as the variable?

25. A compound microscope has two objective lenses. One lens is 10X and the other objective lens is 43X.
    a) If the eyepiece on this microscope is 10X, what is the highest magnification possible for this microscope?
    b) If the image under the microscope is too dark, what part of the microscope must be adjusted?

26. Design a controlled experiment to test the following question.
    "If wind is blowing against a plant, will this cause the plant to grow in the opposite direction?"
    Be sure you:
    a) State a hypothesis.
    b) Explain how you would design an experiment to test that hypothesis.
    c) Explain what important data would have to be collected during that experiment.

Answer Key

Answers Explained Section 1-1

1. Answer for a.

Step 1: Possible hypothesis—Colorful male birds do attract females of the same species more successfully than less colorful male birds.

Step 2: Possible experiment—Put a colorful male bird and a male bird with less color in an aviary with several female birds of the same species.

Step 3: Collect data—Record observations of the female birds and how they interact and attempt to mate with each of the male birds placed into the aviary.

Step 4: Analysis of data and conclusion—If the observational data shows that the colorful male bird attracts more female birds than the less colorful male bird, and then the hypothesis is backed up by the data. If the observational data shows the less colorful male bird attracts more female birds than the colorful male bird, the hypothesis is not validated and a new hypothesis must be created.

Step 5: Is the experiment repeatable?—The results of the experiment must be repeatable for true validity to be obtained.

Answer for b.

Step 1: Possible hypothesis—Carpenter ants prefer oak over pinewood to make their nests in.

Step 2: Possible experiment—Place a carpenter ant colony into a glass tank that has both oak and pinewood inside.

Step 3: Collect data—Record observational data of what wood is chosen to construct the colony nest.

Step 4: Analysis of data and conclusion—If the observational data recorded shows that the carpenter ants choose the oak wood to build their nest, then the hypothesis is validated. If the observational data recorded shows that the carpenter ants choose the pinewood to build their nest, then the hypothesis is not validated and a new hypothesis must be created.

Step 5: Is the experiment repeatable?—The results of the experiment must be repeatable for true validity to be obtained.

Answer for c.

Step 1: Possible hypothesis—Rose bushes grow better in acidic soil compared to alkaline soil.

Step 2: Possible experiment—Plant five rose bushes in acidic soil, five rose bushes in alkaline soil, and five rose bushes in neutral soil as a control group.

Step 3: Collect data—After two weeks, check all three groups of rose bushes to see which group has shown the best growth.

Step 4: Analysis of data and conclusion—If the experimental data shows the rose bushes planted in acidic soil grew the best, the hypothesis is validated. If the experimental data shows the rose bushes planted in the alkaline soil or neutral soil grew better, then the hypothesis is not validated and a new hypothesis must be created.

Step 5: Is the experiment repeatable?—The results of the experiment must be repeatable for true validity to be obtained.

2.

4,000 millimeters	=	4	meters
35.6 kilometers	=	35,600	meters
77° Fahrenheit	=	25	Celsius
287 milliliters	=	0.287	liters
599 grams	=	0.599	kilograms
4 liters	=	4,000	cubic centimeters

3. A hypothesis is a testable predication about what you think will happen during your testing experiment. A hypothesis is a statement of belief concerning the solution to a problem. A theory, on the other hand, is also an explanation for a proposed problem, but this explanation has been backed up by a good deal of data from many experiments. A theory is a well established principle.

4. Some possible hypotheses to this problem include: 1) The TV does not turn on because the batteries in the remote are dead. 2) The TV does not turn on because the TV is not plugged in. 3) The TV does not turn on because the remote control was not pointed properly at the TV.

Answers Explained Section 1-2

1. Story #1—Peter should use the equipment from Group A to complete the protein test. Ninhydrin is added to the test tube with the substance to be tested and then the mixture is heated until bubbles rise from the solution. A color change to purple will indicate protein.

Story #2—Mary should use the equipment from Group E to complete this lab dissection. The earthworm would be pinned to the dissecting pan with the dissecting pins and then the scalpel would be used to open the body of the earthworm. The probe could then be used to remove the intestines from the specimen.

Story #3—Billy should use the equipment from Group F to complete the measurements of the block. The meter stick would be used to measure the width, length, and height of the block. These three measurements are then multiplied to calculate the volume of the block. The block is then placed on the weighing boat and the electronic balance is used to calculate its mass.

Story #4—Sarah should use the equipment from Group C to complete the viewing of this cell sample. The onion sample would be placed on the glass slide, stained with the methylene blue, and then covered with the cover slip. This slide is then ready to be placed under the microscope for viewing.

2. A compound microscope is a scope that magnifies the image you are looking at with two lenses, the ocular lens or eyepiece and the objective lens. The final magnification is the product of the power of the eyepiece lens and the objective lens.

Answers Explained Section 1-3

1. a. Regulation: The earthworm is regulating its body temperature and humidity level.
   b. Nutrition: A caterpillar feeding on a leaf is a type of nutrition.
   c. Transport: The movement of food through the bloodstream of the body is an example of transport.
   d. Regulation: A person screaming as a reaction to a scary image is an example of regulation. The nervous system and endocrine system are activated in a defense mode.
   e. Synthesis: Smaller chemicals combined to make larger more complex chemicals is an example of synthesis.
   f. Growth: The increase in cell number and size within a multicellular organism is an example of growth.
2. Life processes are used to identify something as living or non-living. Scientists have determined that all living organisms perform common actions. These seven common actions that are connected to all living things are called life processes. The seven life processes include: nutrition, transport, respiration, excretion, regulation, growth, and synthesis.

3. Metabolism is all the chemical reactions needed to sustain life working together within the organism. Homeostasis is the ability of an organism to successfully maintain metabolism. When homeostasis is not maintained disease and eventually death can occur.

4. Reproduction is when an organism creates more of the same organism. Reproduction is very critical for the survival of the organism into the future, but is not necessary for the survival of the individual organism. In other words, homeostasis can be maintained without reproduction.

Answers Explained Chapter 1 Exam

Matching Column

1. C	4. F	7. B	10. C	13. A
2. D	5. G	8. A	11. E	14. G
3. E	6. H	9. D	12. B	15. F

Multiple Choice

16. **C** is the correct choice because if you counted the number of petals on a flower, the information collected would be data that contains numbers. Quantitative data contains numbers.

    A is not correct because the data collected to describe the color of a flower would not contain numbers.

    B is not correct because the data collected to describe the smell of a flower would not contain numbers.

17. **D** is the correct choice because a scientist would be able to test the sugar levels of a person after they had eaten. Also, a hypothesis should be a statement predicting the outcome of a particular event.

    A is not correct because it is question. A hypothesis is not a question but a prediction statement made about a particular matter. The question in choice A could lead to a possible hypothesis, but as it is would not be considered a hypothesis.

    B is not correct because if the planet is as of yet undiscovered, the scientist will not be able to test if the hypothesis is correct. A hypothesis that cannot be tested is not a good hypothesis.

    C is not correct because the statement given is an opinion. Opinions like the one stated in choice C cannot be proven correct or incorrect. The conclusion of this statement would lead to different answers based on who tested the hypothesis.

18. **C** is the correct choice. To convert 2,000 micrometers into millimeters you would use the math proportion shown here.

$$\frac{1{,}000 \text{ micrometers}}{1 \text{ millimeter}} = \frac{2{,}000 \text{ micrometers}}{x \text{ millimeters}}$$

You cross multiply the proportion to get the following equation:

1,000 micrometers (x) = 2,000 micrometers (1 millimeter)

When you solve for the (x) in the equation you get the number 2.

A, **B**, and **D** are incorrect for the math reasons stated above.

19. **B** is the correct choice because only the electron microscope has a magnification power high enough to be able to view a virus. Most viruses are very small, about 100 nanometers (one billionth of a meter).

A is incorrect because even a high magnification compound microscope will not be able to view something as small as a virus.

C is not used for magnification, but for separating objects based on density.

D is not used for magnification, but for measuring the mass of objects.

20. **C** is the correct choice because Benedict's solution is used to test for the presence of sugars in a chemical sample.

A is not correct because ninhydrin is used to test for the presence of proteins.

B is not correct because Lugol's iodine is used to test for the presence of starches.

D is not correct because methylene blue is used to stain objects for better visibility under the microscope.

21. **A** is the correct choice because the definition of synthesis is the joining of small molecules (such as amino acids) to create a larger more complex molecule (such as protein).

B is not correct because transport is the movement of materials throughout the cell and in multicellular organisms the movement of materials throughout the body.

C is not correct because respiration is the oxidation of food for energy release.

D is not correct because nutrition is the obtaining and use of food.

22. **B** is correct because regulation is the actions that permit an organism to maintain homeostasis. In this particular example the turtle is trying to maintain a certain body temperature and achieves this goal by sunning itself on the log.

A is not correct because growth is the increase in the cell size or the number of cells within an organism.

C is not correct because respiration is the oxidation of food for energy release.

D is not correct because nutrition is the obtaining and use of food.

Short Response

23. The doctor has designed an experiment to test the effectiveness of a new weight loss pill. Important in any experiment is to include a control. The control is used as a basis of comparison against that which is being tested (variable). In this particular experiment, the 100 people who are not given the pill are considered the "control group." Only if the people who take the pill lose weight and the people who do not take the pill stay at the same weight can the pill be considered a possible method of weight loss.

24. The variable in this experiment would be the temperature at which the lima beans are grown. This can be identified as the variable because it is the only condition that changes between each experimental set-up.
25. Remember, when someone looks through a compound microscope they are looking through two lenses. Compound microscopes have one eyepiece lens, and then a choice of usually two or three objective lenses. But only one objective lens can be used with the eyepiece at a time. So with this particular microscope if you want the highest possible magnification it would be the 10X eyepiece multiplied to the highest object lens that is 43X, giving a total magnification power of 430X.

    If the image under the microscope is too dark, the diaphragm on the compound microscope can be adjusted to increase or decrease the amount of light shining through the specimen on the stage.
26. The following is one possible experiment that could be conducted to test that question.

    Hypothesis: Wind will cause the plant to grow in the opposite direction.

    Experiment design: Grow two plants of the same species under the same conditions, except (the variable) have a constant wind blowing against one plant during the time of the experiment.

    Data collection: During the experiment, data would be collected concerning the direction of growth for each plant.

Chemistry of Life

All living things are made of chemicals. In order to best understand living things and life itself, one must have some working knowledge of the chemistry involved. This chapter begins with some instruction about basic chemistry and the laws that govern this field of science. The chapter then focuses on the branch of chemistry that deals more specifically with the chemicals necessary for life, called organic chemistry. The chapter ends with an overview of enzymes and their importance to all the chemistry that goes on inside of living organisms.

Lesson 2-1: Basic Chemistry

A living organism can be compared to a bag full of varying chemical reactions. All living organisms on Earth perform life processes that involve many complex chemical reactions. To understand these processes of life, it is critical to have at least some basic knowledge of the world of chemistry. It is becoming more and more apparent in the world of biology that a true understanding of life requires an understanding of chemistry. This chapter should help in your understanding of this topic.

The Building Blocks of Matter

All living things are made of matter. The smallest block of matter is the **atom**. Atoms are not alive. An atom is the smallest particle that retains the properties of that particular piece of matter. For example, if I was to break apart a sample of pure silver metal into increasing smaller pieces, I would eventually have the smallest piece of silver possible, a silver atom. I cannot break a silver atom down in anything smaller and still have silver. So what would I have if I were to break that atom of silver down even further?

If an atom is broken up, the properties of that atom also disappear. The number and types of particles that make up an atom determine the properties of the particular atom. Inside most atoms you will find three important subatomic particles: protons, neutrons, and electrons.

Protons are positively charged particles found at the center or the nucleus of the atom. **Neutrons** are particles that have no charge and are also found in the nucleus of the atom. The protons and neutrons are held together with very strong forces called nuclear forces. Nuclear energy is the result of breaking the bonds that hold these subatomic particles in the nucleus together. **Electrons** are negatively charged particles that are found moving at high speeds around the nucleus of the atom in an area called the ***electron cloud***. The electron cloud is basically a way for chemists to say they know the electrons are in a general area around the nucleus, but at any one moment they do not know exactly what location in that particular region. The negatively charged electrons are attracted to the positively charged protons in the nucleus, but the high-speed movement of the electrons keeps them from being pulled into the positively charged nucleus. A good way to try to picture this is to think of a rock tied to a piece of string. The rock is drawn to the earth by gravity, but if someone were to spin the string in a circle with the rock attached, the rock would resist the force of gravity as long as the rock was kept spinning on the string. Table 2.1 shows some other important information about the characteristics of these three subatomic particles with which you should become familiar.

Comparison of Major Subatomic Particles

Subatomic Particle	Charge	Mass	Location
Proton	+	1*	Nucleus of the atom
Neutron	0	1*	Nucleus of the atom
Electron	–	1/1837 the mass of a proton	Electron cloud outside the nucleus of the atom

* 1 atomic mass unit = 1/12 the mass of a carbon-12 atom

Table 2.1

Depending on what kind of atom is broken up, you will find different numbers of these three subatomic particles. The number of protons inside the atom, sometimes called the **atomic number**, determines the type of atom.

Example

If you know an atom contains six protons, how can you determine the type of atom?

Answer:

If you look in the periodic table of elements, the only atom that contains six protons (atomic number of six) is carbon. Six is the identifying number of protons for an atom of carbon, and no different atom would have this number of protons.

••••••

As soon as you know the number of protons inside the nucleus of the atom, you can determine the identity of the atom. A substance in nature that is made up of only one particular kind of atom is called an **element**. Table 2.2 shows the subatomic particle arrangement of some of the most common elements found in living organisms.

Subatomic Particles Found in Some Elements Common in Living Organisms

Element	Protons	Neutrons	Electrons
Carbon	6	6	6
Oxygen	8	8	8
Hydrogen	1	0	1
Nitrogen	7	7	7

Table 2.2

Let us suppose you have a pure sample of the element gold. If you physically broke this sample of gold into the smallest gold pieces possible, all of the pieces would be individual atoms of gold. Each of these individual atoms of gold would contain 79 protons. If these gold atoms were broken down any further, they would no longer be gold atoms and have the characteristics of gold. You would basically be left with many random protons, neutrons, and electrons that had once created gold atoms.

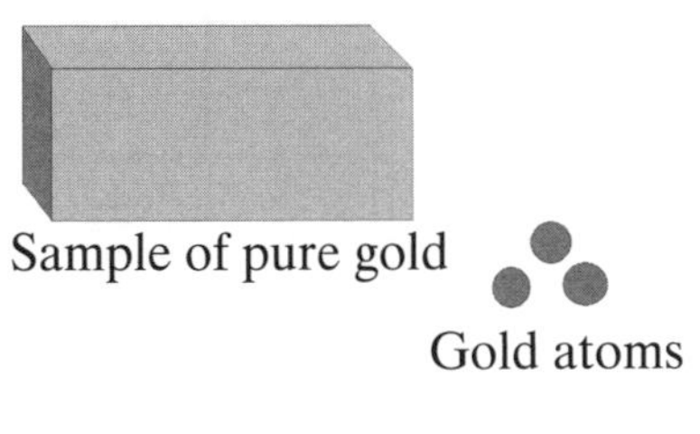

There are 92 naturally occurring elements on this planet. These elements are arranged in atomic number order from smallest to largest in a chart used by most chemists and biologists called the **periodic table**.

This means there are 92 elements on Earth that were not created in a chemistry lab by a scientist. In recent history, scientists have been able to create a number of unstable, short-lived elements, bringing the total number of elements to about 118. In the periodic table, the elements are represented by symbols. Symbols are the shorthand way of showing the presence of a particle element. Many of these symbols are based on the Latin names for these elements.

The Periodic Table of Elements

Hydrogen 1.0079 H 1																	Helium 4.003 He 2
Lithium 6.941 Li 3	Beryllium 9.012 Be 4											Boron 10.811 B 5	Carbon 12.011 C 6	Nitrogen 14.007 N 7	Oxygen 15.999 O 8	Fluorine 18.998 F 9	Neon 20.180 Ne 10
Sodium 22.990 Na 11	Magnesium 24.305 Mg 12											Aluminum 26.982 Al 13	Silicon 28.086 Si 14	Phosphorus 30.974 P 15	Sulfur 32.065 S 16	Chlorine 35.453 Cl 17	Argon 39.948 Ar 18
Potassium 39.098 K 19	Calcium 40.078 Ca 20	Scandium 44.956 Sc 21	Titanium 47.867 Ti 22	Vanadium 50.942 V 23	Chromium 51.996 Cr 24	Manganese 54.938 Mn 25	Iron 55.845 Fe 26	Cobalt 58.933 Co 27	Nickel 58.693 Ni 28	Copper 63.546 Cu 29	Zinc 65.39 Zn 30	Gallium 69.723 Ga 31	Germanium 72.64 Ge 32	Arsenic 74.922 As 33	Selenium 78.96 Se 34	Bromine 79.904 Br 35	Krypton 83.80 Kr 36
Rubidium 85.468 Rb 37	Strontium 87.62 Sr 38	Yttrium 88.906 Y 39	Zirconium 91.224 Zr 40	Niobium 92.906 Nb 41	Molybdenum 95.94 Mo 42	Technetium {98} Tc 43	Ruthenium 101.07 Ru 44	Rhodium 102.906 Rh 45	Palladium 106.42 Pd 46	Silver 107.868 Ag 47	Cadmium 112.411 Cd 48	Indium 114.818 In 49	Tin 118.710 Sn 50	Antimony 121.760 Sb 51	Tellurium 127.60 Te 52	Iodine 126.904 I 53	Xenon 131.293 Xe 54
Cesium 132.905 Cs 55	Barium 137.327 Ba 56	Lutetium 174.967 Lu 71	Hafnium 178.49 Hf 72	Tantalum 180.948 Ta 73	Tungsten 183.84 W 74	Rhenium 186.207 Re 75	Osmium 190.23 Os 76	Iridium 192.217 Ir 77	Platinum 195.078 Pt 78	Gold 196.967 Au 79	Mercury 200.59 Hg 80	Thallium 204.383 Tl 81	Lead 207.2 Pb 82	Bismuth 208.980 Bi 83	Polonium {209} Po 84	Astatine {210} At 85	Radon {222} Rn 86
Francium {223} Fr 87	Radium {226} Ra 88	Lawrencium {262} Lr 103	Rutherfordium {261} Rf 104	Dubnium {262} Db 105	Seaborgium {266} Sg 106	Bohrium {264} Bh 107	Hassium {277} Hs 108	Meitnerium {268} Mt 109	Darmstadtium {281} Ds 110	Unununium {272} Uuu 111	Ununbium {285} Uub 112	113	Ununquadium {289} Uuq 114	115	Ununhexium {292} Uuh 116	117	118

Lanthanum 138.906 La 57	Cerium 140.116 Ce 58	Praseodymium 140.908 Pr 59	Neodymium 144.24 Nd 60	Promethium {145} Pm 61	Samarium 150.36 Sm 62	Europium 151.964 Eu 63	Gadolinium 157.25 Gd 64	Terbium 158.925 Tb 65	Dysprosium 162.50 Dy 66	Holmium 164.930 Ho 67	Erbium 167.259 Er 68	Thulium 168.934 Tm 69	Ytterbium 173.04 Yb 70
Actinium {227} Ac 89	Thorium 232.038 Th 90	Protactinium 231.036 Pa 91	Uranium 238.029 U 92	Neptunium {237} Np 93	Plutonium {244} Pu 94	Americium {243} Am 95	Curium {247} Cm 96	Berkelium {247} Bk 97	Californium {251} Cf 98	Einsteinium {252} Es 99	Fermium {257} Fm 100	Mendelevium {258} Md 101	Nobelium {259} No 102

Example

The symbol for the element iron is **Fe**.

This symbol is derived from the Latin word for iron, which is *ferrum*.

••••••

Although the periodic table is full of many elements found on this planet, only a few become very important in the study of biology. Approximately 97 percent of most living organisms are made up of only six elements—hydrogen, carbon, oxygen, nitrogen, sulfur, and phosphorus.

Practice Section 2-1

1. What is the name of the subatomic particle that has a positive charge and is found in the nucleus of the atom?

   a) proton   b) neutron   c) electron   d) isotope

2. The atomic number of an element is six. What is the number of protons in the nucleus of that atom?

   a) 3   b) 6   c) 12   d) 18

3. If an atom has three electrons, three neutrons, and three protons, what is the charge on the nucleus of that atom?

   a) 0   b) 3   c) 6   d) 9

4. Which of the following has the most mass?

   a) 2 protons   c) 1,000 electrons

   b) 3 neutrons   d) they all have the same mass

5. If you have a sample of a particular element, what is the smallest piece of that element you can have that will retain the particular properties of that element?

   a) proton   b) atom   c) electron   d) neutron

6. There are about 92 naturally occurring elements found on this planet. All of these elements can be represented by symbols. What is the symbol for carbon?

   a) Ca   b) Cn   c) C   d) Cb

7. What subatomic particle identifies an atom as that of a particular element?

8. What are the subatomic particles found in the nucleus of an atom? What is the charge on the nucleus?

9. Explain what keeps the electrons confined to the space surrounding the nucleus of an atom.

Lesson 2-2: Chemical Bonding

Very few of the 92 elements found in the periodic table would be found as pure samples in the natural environment. Many of these 92 elements are chemically combined with other elements to form what are called compounds. **Compounds** are created when you join two or more elements chemically. In so doing, these elements in the compound now behave differently then the elements would have if they were alone.

An excellent example of this behavior of elements in compounds is illustrated here.

- A pure sample of the element sodium (Na) is a very volatile, explosive, gray metal.
- A pure sample of the element chlorine (Cl) is a very poisonous, green gas.

But when you chemically combine one atom of chlorine with one atom of sodium, you create NaCl (table salt). The original properties that each element had alone give way to the very new properties these elements have now chemically combined.

••••••

Chemical Reactions and Bonding

Chemical bonds are created when two or more atoms are combined chemically. When a compound is chemically broken down, it is the chemical bonds that are broken. All of chemistry involves the study of bonds forming and breaking between atoms within compounds. A chemical reaction is the process of chemically joining elements to make compounds or chemically breaking compounds to make elements or new compounds. All living organisms have millions of chemical reactions occurring inside of them all the time. These chemical reactions enable organisms to achieve the life processes necessary to remain alive, such as processing food, removing waste, regulating the internal environment, and producing more of its own kind.

Electron Shells

Atoms, as mentioned earlier, have specific numbers of electrons orbiting around their nuclei. The number of electrons that are found in a given area around the nucleus is based on the amount of energy those particular electrons may have. The space that these electrons can orbit has a very specific arrangement. Closest to the nucleus is an area that can hold a maximum of two electrons and is sometimes called the first shell. The second energy shell around the nucleus can hold a maximum of eight electrons, and the third shell also holds a maximum of eight. Atoms are said to be chemically stable if the last shell, also called the **valence shell**, circling the atom is filled with electrons.

Atoms with full valence shells usually do not engage in chemical reactions with other atoms, while on the other hand atoms that do not have the last shell filled with electrons are structurally unstable and look to engage in chemical reactions with other elements and form bonds with them. The bonds that are formed usually make the elements more chemically stable, because in forming the bonds the last electron shell will become full.

In most atoms the number of protons and electrons are equal. This means that the atom has no net charge.

Example

An atom of carbon has six protons and six electrons, six positive charges added to six negative charges equals zero.

6 protons = 6^+ and 6 electrons = 6^-, so $6^+ + 6^- = 0$

This means that the total net charge on that atom is zero.

• • • • • •

Chemical Bonding

Sometimes, when atoms get close together, one atom loses one or more electrons to the other atom. This means that the ratio of electrons to protons is no longer equal. Both of these atoms no longer have a net charge of zero. The atom that has gained the electron or electrons will take on a negative charge, while the atom that lost the electron or electrons will take on a positive charge. An atom that has a positive or negative charge is no longer called an atom, but instead called an **ion**. When two atoms are involved in an electron exchange where they become ions, the charges of these ions hold the two atoms together like opposite sides of a magnet. The bond holding these ions together is called an **ionic bond**. The following is an illustration of ionic bonding between a sodium atom and a chlorine atom to produce the ionic compound called sodium chloride, or table salt.

Ionic Bonding

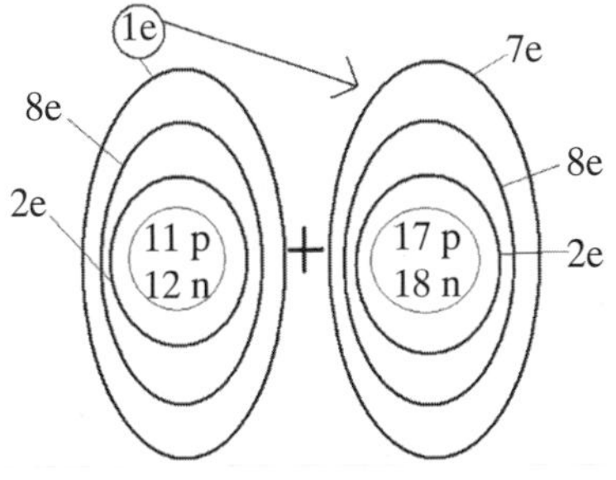

Sodium atom + Chlorine atom

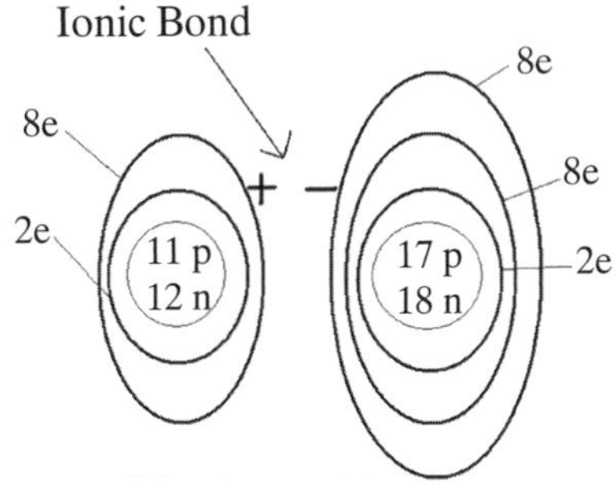

Sodium ion + Chlorine ion

When two atoms draw close together, sometimes neither atom wants to lose an electron. In this case, each atom will share some of its electrons with the other atom. Sharing electrons means that the electrons in question are moving around the nuclei of both atoms. In the act of sharing the electrons, the electron clouds of the two atoms overlap. The sharing of the electrons in this case is the bond that holds the atoms together. It may be helpful to think of two people who want a broomstick. If each person has his hand on the broomstick and won't let go, wherever one person goes, the other must follow. The broomstick is the bond that holds them together. It is similar in the case of two atoms sharing or holding on to some electrons. The bond formed by shared electrons is called a **covalent bond**. When atoms are joined by covalent bonds, the compound they form is called a *molecule*. A molecule is the smallest unit of most compounds that are studied in biology. The following is an illustration of one atom of oxygen covalently bonded to two atoms of hydrogen to produce one molecule of water.

Covalent Bonding in a Water Molecule

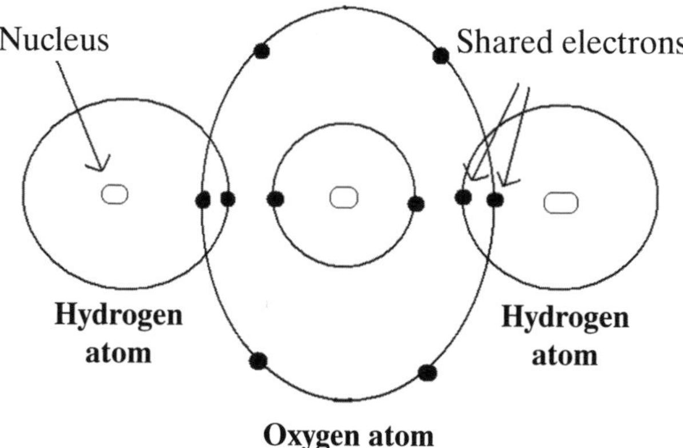

Sometimes the electrons are shared between two atoms equally and, sometimes, because of the nature of each atom, the electrons being shared spend more time with one atom than the other. When two atoms share electrons unequally, the bond that is formed between them is called a ***polar covalent bond***. When electrons are shared equally to create a bond, the bond that is formed is called a ***nonpolar covalent bond***. One very important example of polar covalent bonds is found in the structure of water. The electrons in this compound are held closer to the oxygen atom than the two hydrogen atoms. The oxygen is said to have a higher **electonegativity** than the hydrogen atoms, meaning oxygen has a greater attractive force for electrons than hydrogen. Because the electrons are being shared, but remain closer to the oxygen atom during the sharing, the oxygen atom acquires a slightly negative charge. At the same time, because of the condition of the electrons being shared unequally, the hydrogen atoms acquire a slightly positive charge.

Intermolecular bonds

Covalent and Ionic bonds are types of intramolecular bonds. Intramolecular bonds are bonds that hold atoms togther to create compounds or molecules.

Hydrogen Bonding Between Water Molecules

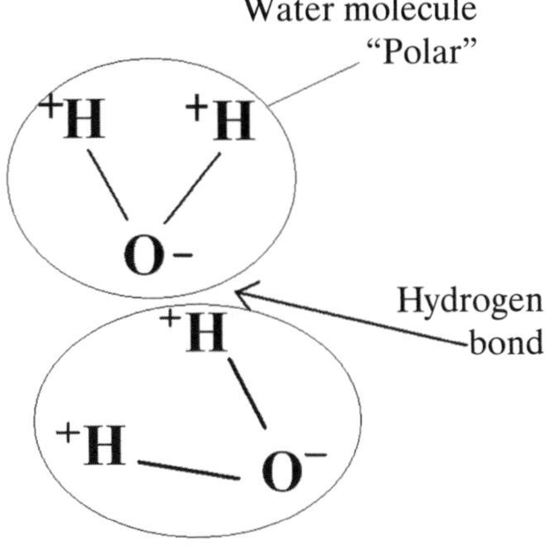

On the other hand, a bond that holds two or more molecules together is called an intermolecular bond. A molecule is considered to be polar if the charges of the atoms in the molecule are not distributed evenly. Polar molecules display weak attractive forces for other polar molecules. The positive side of one molecule attracts the negative side of a different polar molecule.

For instance, a water molecule contains a slightly positive hydrogen atom on one side of the compound. This hydrogen atom will attract negatively charged particles from other compounds. The bond that forms between a slightly positive hydrogen atom in one molecule and a negatively charged atom in another molecule is called **hydrogen bonding**. A hydrogen bond is an example of a type of intermolecular polar bond that involves the element hydrogen in the actual bond.

The polar nature of water is very important in the study of biology. Living organisms are composed mostly of water. Many substances that are necessary for life, such as most food molecules, are also polar in nature. This means that these polar substances will easily dissolve into the polar medium of water because of the chemical nature of these materials.

In ***nonpolar*** substances, there is no separation of positive and negative charges in the molecule. So, in essence, the entire molecule behaves like there is no charge. For example, if there are three molecules close together, two being polar and one being nonpolar, the two polar molecules will move together and the nonpolar molecule will move away. You can see an example of this chemical process by pouring some vegetable oil into a glass of water. The nonpolar vegetable oil will not mix with the polar water. On a

An Illustration of the Separation of Polar and Nonpolar Molecules

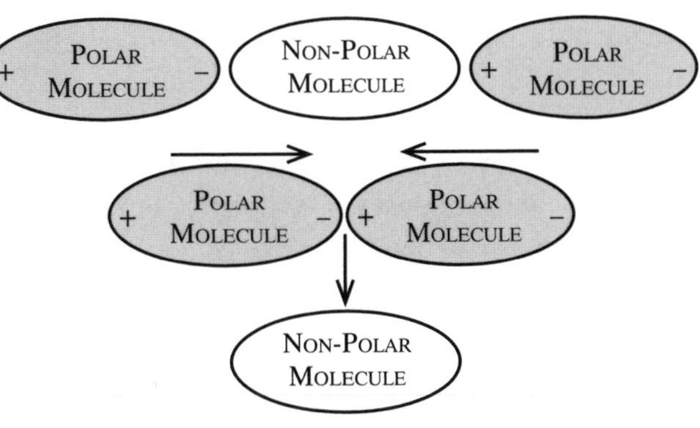

molecular level, what is occurring is that the unlike charges found in each of the two polar molecules are attracted to one another. These attractive forces bring the two polar molecules together. The nonpolar molecule is not attracted to either of the polar molecules and is pushed aside when the polar molecules are drawn together.

If you were then to pour in a polar substance such as table salt, it would dissolve quickly into the polar water. This happens because all of the molecules are attracted to one another on a molecular level.

Bonding of Water Molecules

As previously demonstrated, water can form hydrogen bonds with other molecules. In fact, each water molecule can form up to four hydrogen bonds with four separate molecules. Even though these hydrogen bonds are not as strong as covalent or ionic bonds, they give water some special properties.

Cohesion is the chemical property of two like molecules bonding together. When water forms hydrogen bonds with other water molecules, these are said to be cohesive forces holding the molecules together. Water molecules are very cohesive, which makes water somewhat "sticky" with other water. Many organisms on this planet use this "sticky" property of water to their advantage. For instance, some insects are able to swim on the surface of water because of the cohesive forces of the water molecules below them.

Adhesion is the attractive forces between two molecules that are different. Water also can be "sticky" to other molecules. For example, many plants transport water through their stems in low narrow tubes. In these tubes, the forces of cohesion, water bonding with water, and adhesion, water bonding to the sides of these tubes, help the flow of water up the tube to the top of the plant. This action of water moving up a narrow tube by adhesion and cohesion is called ***capillary action.*** The following illustration shows that as the tube becomes narrower, the height that the water will climb increases. The height of the water increases when the tube is narrower because the adhesive forces of the molecules increases. Capillary action will only work up to a given height; eventually, the force of gravity becomes stronger than these forces.

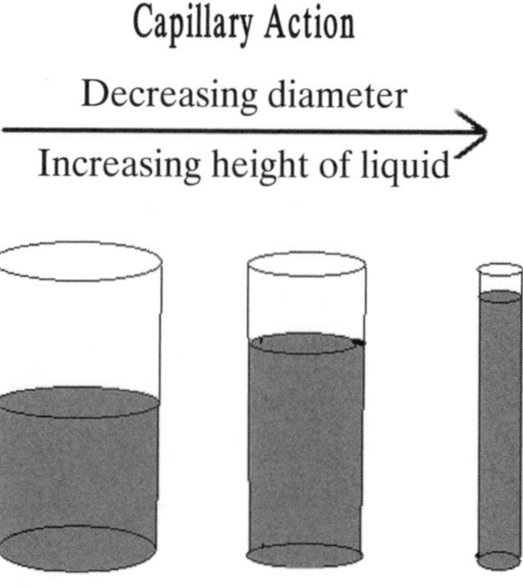

All living things are filled with chemicals undergoing many different chemical reactions. Most of these chemicals found within living organisms are dissolved in the chemical water found within their cells. The chemicals that are being dissolved are called **solutes**, while the chemical that does the dissolving is called a **solvent**. Most organisms on this planet have large quantities of water inside of their bodies. One reason that most living things contain water is that water is a very good solvent. This means that many substances on this planet that are necessary for life will dissolve in water, allowing the needed chemical reactions of life to occur freely.

Mixtures

When different types of molecules are put together, but are not joined chemically, you have what a chemist would call a **mixture**. The chemicals in a mixture all retain their original properties. Mixtures are important in many conditions associated with living things. For example, the air we breathe is a mixture. It is good for us that the air on this planet is a mixture. Air contains the element oxygen, which is needed for respiration. If air was not a mixture, we would not be able to extract this element from the atmosphere by breathing. Table 2.3 shows the percentages of materials found in the mixture of a sample of air.

The Percentage of Gases Found in Air

Gas	Percentage by Volume
Nitrogen	78.1
Oxygen	20.9
Argon	0.93
Carbon dioxide	0.035

Table 2.3

Water vapor is also present in the air in amounts ranging from 0% to 5%. Because air is a mixture humans can extract oxygen from it during breathing for the chemical process of respiration.

pH and Life

When a compound such as table salt is placed into water, the ions of sodium (Na) and the ions of chlorine (Cl) that make up the ionic compound will break apart. The polar nature of water and of salt causes the mixing of these substances. The positive sodium ions stick to the negative sides of the polar water molecules, while the negative chlorine atoms stick to the positive sides of the water molecules.

Most water molecules do not break up into individual elements. But some molecules of water can, at times, break apart into the ions (H^+) and (OH^-), even though water is not an ionic compound.

Ionization of Water

$$H_2O \rightleftharpoons H^+ + OH^-$$

Water — Hydrogen ion — Hydroxide ion

The amount of H^+ and OH^- ions in a solution can be measured on what is called a ***pH scale***. If the levels of H^+ ions are higher than the levels of OH^- ions in a solution, then the substance is called an **acid**. If the levels of OH^- ions are higher than the levels of H^+, then the substance is called a **base**. On the pH scale, the numbers run from 0 to 14, with the number 7 signifying that the levels of H^+ and OH^- ions are equal (water). The numbers from 7 down to 0 are used to categorize ever stronger acids, while the numbers from 7 to 14 are used to show stronger and stronger bases. Each unit on the pH scale is equivalent to a 10 times difference in strength of the acid or base.

Example

A substance with a pH of 4 is an acid.

A substance with a pH of 2 is an acid.

The substance with the pH of 2 is 100 times stronger than the substance with a pH of 4.

The substance with a pH of 2 is 2 units away from the substance with a pH of 4 on the pH scale. Each unit on the pH scale is valued at 10. So the math is $10 \times 10 = 100$.

Conversely:

A substance with a pH of 9 is a base.

A substance with a pH of 12 is a base.

The substance with a pH of 12 is 1,000 times stronger than the substance with a pH of 9 on the pH scale. Each unit on the pH scale is valued at 10. So the math is $10 \times 10 \times 10 = 1{,}000$.

••••••

Practice Section 2-2

1. What is the maximum number of electrons that can be found in the second energy shell of an atom?
   a) 2 b) 4 c) 6 d) 8
2. In covalent bonds, the electrons are being
   a) lost b) gained c) shared d) A and B
3. Which of the following pH values represents the strongest base?
   a) 4 b) 8 c) 12 d) 14
4. All of the following are elements except
   a) carbon b) oxygen c) water d) nitrogen
5. Which of the following subatomic particles is located outside the nucleus?
   a) proton b) electron c) neutron d) A, B, and C
6. After taking a shower you notice that some water droplets are clinging to the shower curtain. This is an example of
   a) cohesion
   b) adhesion
   c) ionic bonding
   d) dehydration synthesis

7. What is a chemical bond?
8. How do ions form?
9. Is water a polar or non-polar molecule? How does this characteristic of water become important in this compounds ability to function as a solvent?

Lesson 2-3: Organic Chemistry

Living things are made of many different elements. Many of the elements found in living things are also found in nonliving things, but there are certain elements found in greater percentages in living organisms. The following is a list of these elements and at least some of their functions in living organisms.

Elements That Are Found in the Greatest Percentages in Living Matter

Carbon	The major element found in all organic molecules.
Hydrogen	A major element found in all organic molecules and part of the structure of water.
Oxygen	A major element found in many organic molecules and part of the structure of water.
Nitrogen	A major element found in all proteins and nucleic acids.
Sulfur	Important element found in most proteins.
Phosphorus	The high-energy bond connects to this element in all ATP molecules.
Magnesium	Part of the chlorophyll molecules in photosynthetic organisms and essential for some enzymes.
Iodine	Important for the creation of the hormone thyroxin.
Iron	Part of the hemoglobin structure in the red blood of many animals.
Calcium	Part of bone structures. Needed for the actions of muscle contractions.
Sodium	Important for the movement of nerve impulses in most animals.
Chlorine	Important element in the creation of digestive chemicals and in the chemical processed of photosynthesis.
Potassium	Important for the movement of nerve impulses in most animals.

Table 2.4

Scientists originally thought that organic chemicals could only be created by a living organism and believed that the chemistry of these materials was totally different than that for inorganic compounds. In the early 1800s, when organic chemicals were created in the lab for the first time, the original ideas about organic compounds had to be rethought.

Today, **organic compounds** usually refer to a group of compounds that contain the elements carbon and hydrogen, and many times other elements such as nitrogen and oxygen. Organic compounds are usually very large molecules that are produced by plants and animals. Carbon atoms have four valence electrons that allow it to form up to four covalent bonds with other elements. This chemical property of carbon enables this element to be formed into long chains and rings. This is the reason why organic compounds can be very large, complex structures.

```
   H
   |
H -C- H
   |
   H
```

Methane

Example

> The organic compound methane can form covalent bonds with four separate elements at the same time.

••••••

The four major organic compounds found and produced in the living cell are: **carbohydrates, proteins, lipids,** and **nucleic acids.**

These compounds are all considered to be *macromolecules* because of their very large size. These four macromolecules are each created by smaller subunits.

Macromolecule	Subunit
Carbohydrates	Sugars (monosaccharide, disaccharide, and polysaccharide)
Proteins	Amino acid
Lipids	Glycerol and Fatty Acids
Nucleic Acids	Nucleotides

Carbohydrates

Carbohydrates are compounds made up of the elements hydrogen and carbon in the ration of 2:1. This means that when you count all the elements in a carbohydrate compound you will have twice as many hydrogen atoms as carbon atoms. A **structural formula** is a diagram that shows all the elements present and how they are connected to one another in a compound. Looking at the structural formula of a simple carbohydrate such as glucose, the lines represent where pairs of electrons are being shared between the atoms. So, each line represents a covalent bond in the compound. In glucose, if you count the number of Hs, which stand for the hydrogen atoms in the compound, and you count the number of Cs, which stand for the carbon atoms in the compound, you will see there are 12 hydrogen atoms and six carbon atoms, a 2:1 ratio as would be expected in a carbohydrate molecule. Counting the number of hydrogen atoms and carbon atoms of an unknown compound is an easy way to identify a carbohydrate when given a structural formula of the molecule.

Glucose

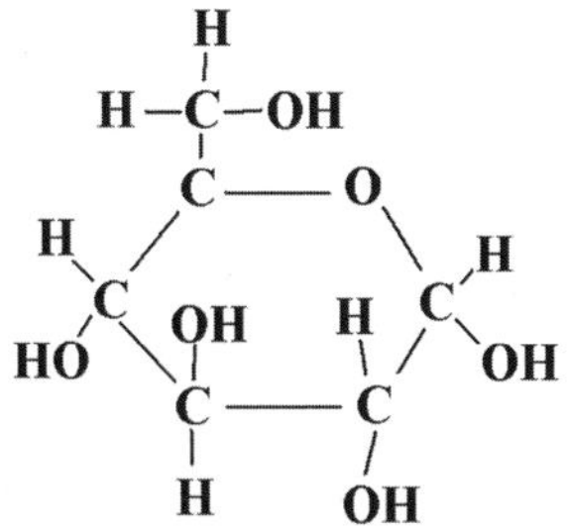

The **molecular formula** for the molecule of glucose just lists the elements present in the compound and the number of atoms of each element present. The molecular formula of glucose is $C_6H_{12}O_6$. Molecular formulas do not provide any clues about the bonding of these atoms.

Disaccharide

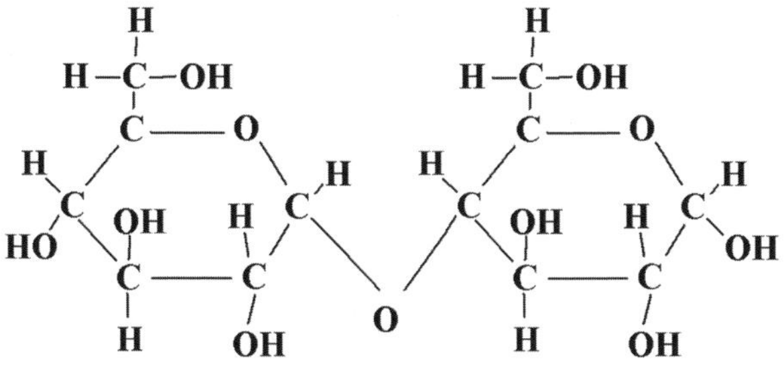

Fructose and ***galactose*** have the same molecular formula but different structural formulas. Compounds that have the same molecular formula but different structural formulas are called ***isomers***. All three of these carbohydrates are called **monosaccharides**. A monosaccharide is a compound that is just one sugar molecule large, thus deemed by biologists and chemists as a ***simple sugar***. If you chemically combine two of these simple sugars, you create a **disaccharide**, or a compound with two sugar molecules. If you combine more than two sugars together you create what is called a complex sugar, or a **polysaccharide**.

Polysaccharides are examples of **polymers**. A polymer is a compound created by joining many smaller identical compounds together in a chain. Starch is an example of a complex sugar, or polymer.

Cells use carbohydrates mainly as a source of energy. In plant cells, the complex sugar called ***cellulose*** is used to support the cell walls. In animals, the complex sugar ***glycogen*** is used as an energy reserve in the body.

General Structure of an Amino Acid

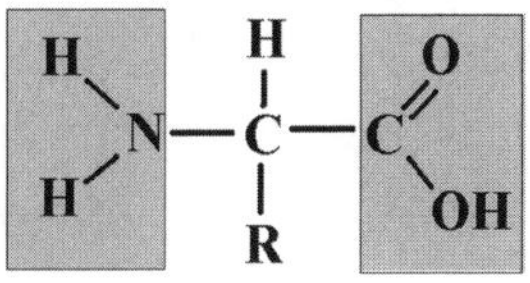

Amino Group

Carboxyl Group

Proteins

A **protein** is a compound that contains carbon, hydrogen, oxygen, nitrogen, and many times also the element sulfur. Proteins are large complex polymers made of the repeating smaller units called **amino acids**. There are 20 different types of amino acids important for living organisms that can be arranged in different polymer formations to form different proteins. In most biology classes it is important to be able to recognize the general structure of the amino acid molecule.

The bond that forms between two amino acid molecules to create a protein molecule is called a ***peptide bond***. A chain of amino acids forms a protein molecule. A chain of chemically combined amino acids is also sometimes referred to as a ***polypeptide***. How the polypeptide is linked together says a lot about the nature of the protein molecule. Proteins can form many different shapes based on how these chains of amino acids twist and connect within the molecule. The shape of the protein determines its function in the organism.

Protein synthesis is a major chemical activity of the cell. Proteins are used in the creation of many of the structures found in living things. Some examples of the important protein compounds formed are ***enzymes***, ***hemoglobin***, and many of the ***hormones*** found in living organisms. Proteins are also important in the formation of the cell itself, because proteins make up part of the cell membrane's structure.

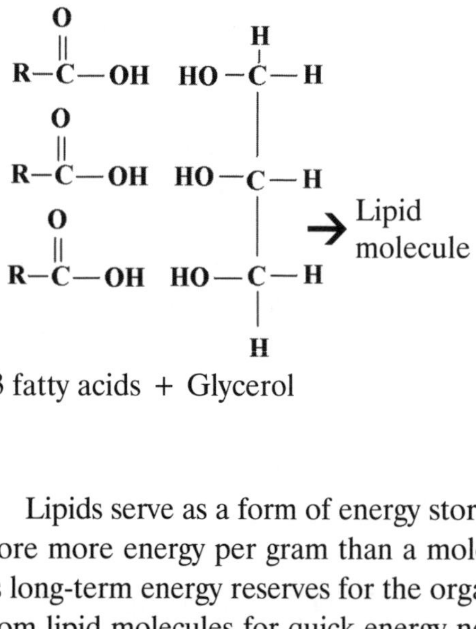

```
   O                  H
   ||                 |
R—C—OH      HO—C—H
   O                  |
   ||                 |
R—C—OH      HO—C—H
   O                  |       → Lipid
   ||                 |         molecule
R—C—OH      HO—C—H
                      |
                      H
3 fatty acids + Glycerol
```

Lipids

Lipids are a group of organic macromolecules that include oils, fats, and waxes. A lipid molecule contains the elements carbon, hydrogen and oxygen. In lipid molecules, the ratio of hydrogen to oxygen is greater than the 2:1 ratio observed in carbohydrates. A glycerol molecule (a type of alcohol) joined with a group of compounds called fatty acids creates most lipid molecules.

Lipids serve as a form of energy storage for cells. A lipid molecule can, in fact, store more energy per gram than a molecule of a carbohydrate. Lipids serve well as long-term energy reserves for the organism. The problem with obtaining energy from lipid molecules for quick energy needs of the cell is that drawing the energy from lipid molecules is slower than and not as easy as drawing energy from carbohydrate molecules. That is why most athletes will have something such as orange slices (high in carbohydrates) during a game, not a bacon sandwich (high in fat).

The creation of carbohydrates, proteins, and lipids from the subunits they are made of is achieved by a chemical process called **dehydration synthesis**. Dehydration synthesis is a reaction where two smaller molecules are joined chemically by removing elements from each of these molecules. The removed elements create a water molecule. An example of the process of dehydration synthesis joining two amino acid molecules is shown on page 53.

Dehydration Synthesis

H H O H H O
N — C — C + N — C — C
H H OH H CH_3 OH

H H O H H O
N — C — C — N — C — C + H_2O
H H CH_3 OH

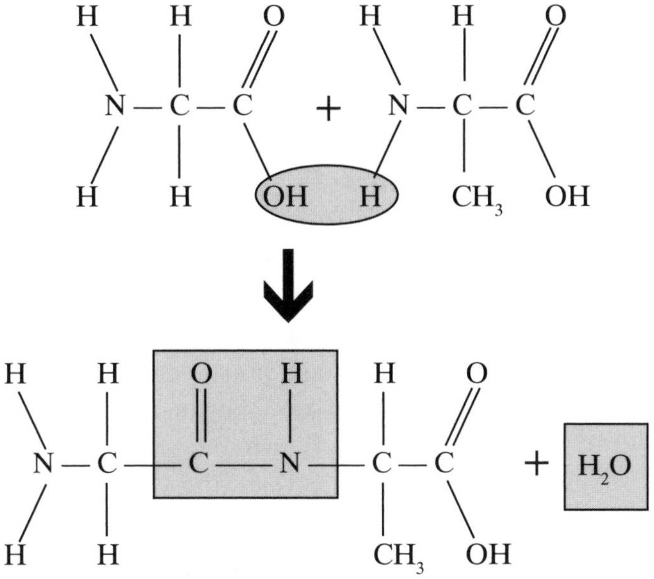

When complex macromolecules such as proteins, carbohydrates, and lipids need to be broken down into simpler substances for energy or for structural parts within an organism, a chemical reaction called **hydrolysis** achieves this process. Hydrolysis is basically the opposite reaction of dehydration synthesis. Where dehydration synthesis joins to smaller chemicals by removing a water molecule,

Hydrolysis

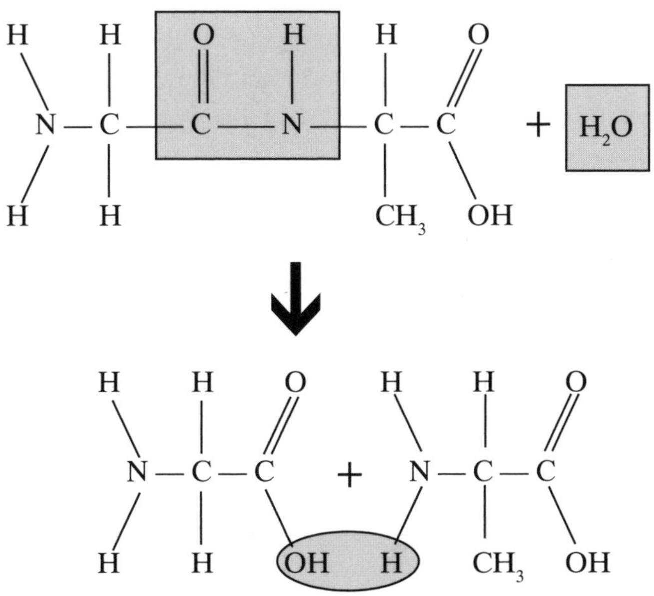

H H O H H O
N — C — C — N — C — C + H_2O
H H CH_3 OH

H H O H H O
N — C — C + N — C — C
H H OH H CH_3 OH

hydrolysis splits apart larger molecules by adding a water molecule. The following is an example of a polypeptide molecule being broken into amino acid molecules by the chemical process of hydrolysis.

Double-Helix Model of DNA

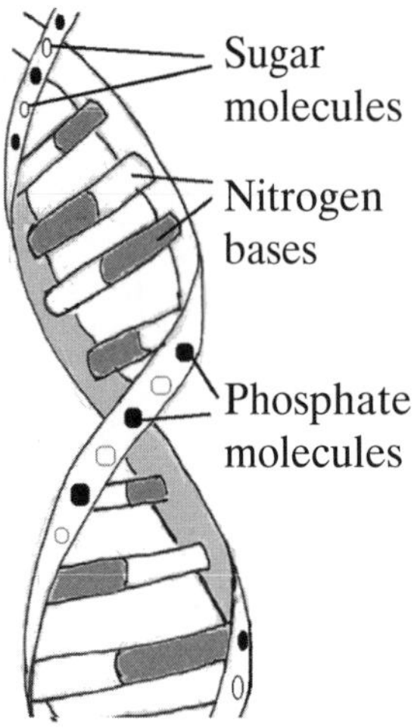

Nucleic Acids

A fourth organic macromolecule important in the study of life is the nucleic acid. Nucleic acids contain the elements carbon, hydrogen, oxygen, nitrogen, and phosphorus. **Nucleic acids** are polymers created by a subunit molecule called the ***nucleotide***.

The sugar molecule that can be found in the nucleotide can either be a ribose sugar or a deoxyribose sugar. The nitrogen bases can be one of the following: guanine, cytosine, thymine, adenine, or uracil. Many repeating nucleotide units create the very famous chemical with the twisted ladder shape called the double helix.

Nucleic acids are the chemicals involved in passing on hereditary information. Two very important nucleic acids are DNA (deoxyribose nucleic acid) and RNA (ribose nucleic acid). Reproduction in all living organisms focuses on the replication of these chemicals and the transfer of these molecules into the future generations. Nucleic acid molecules act as the blueprints for the creation of all life on this planet.

There are some chemicals that are not organic that are important in the study of living organisms. An **inorganic compound** is a compound that does not have both carbon and hydrogen atoms bonded together within the molecule. An example of a very common inorganic compound found in living cells is water. Another common inorganic compound is table salt (NaCl). There are many acids and bases in living organisms that are also inorganic.

Practice Section 2-3

1. Which group of organic molecules is the building block of enzymes?
   a) carbohydrates b) proteins c) lipids d) nucleic acids
2. Which of the following carbohydrates is a polysaccharide?
   a) glucose b) fructose c) starch d) sucrose
3. Which of the following macromolecules is NOT connected properly with its subunit?
   a) simple sugars—carbohydrate
   b) amino acids—protein

c) fatty acids and glycerol—lipids

d) glucose—nucleic acids

4. What element is found in all organic compounds?

   a) nitrogen b) carbon c) iron d) oxygen

5. The formation of a larger molecule from two smaller molecules with the removal of water is called

   a) hydrolysis c) metabolism

   b) dehydration synthesis d) ion formation

6. DNA and RNA are two types of

   a) carbohydrates b) lipids c) polysaccharides d) nucleic acids

7. What is organic chemistry? What are the four major organic compounds produced in living organisms?

8. What is the basic unit of all proteins called? What are the chemical bonds that form proteins called?

9. What are nucleic acids constructed of? What chemicals create that basic unit?

Lesson 2-4: Enzymes and Chemical Reactions

Earlier, living organisms were described as bags of chemical reactions. Indeed, all living things undergo chemical reactions to achieve the processes critical to life. What is important to understand is that all these chemical reactions occur at the right speeds and at the right times because of marvelous organic chemicals called enzymes. **Enzymes** are living chemical catalysts. They are designed to help all living organisms perform the chemical reactions necessary for life. Without enzymes, these chemical reactions would occur at a rate too slow to keep up with the normal metabolism in living organisms.

The Structure of Enzymes

Enzymes are protein based. Enzymes are very complex structures consisting mainly of long chains of amino acids. The largest enzymes are made up of thousands of amino acids. These long chains of amino acids fold into very unique shapes. The shape an amino acid bends into determines the characteristics of the enzyme. There are many different kinds of enzymes in living organisms that control the many different kinds of chemical reactions that occur within an organism. Enzymes are very specific, which means certain enzymes only aid in certain chemical reactions, while other enzymes are particular to other chemical reactions.

Many enzymes require the presence of other compounds called ***cofactors***, or coenzymes, to work. Many times these coenzymes are vitamins. A deficiency in certain vitamins in the body of an organism can cause certain enzymes not to work properly. Enzymes are also sensitive to their chemical surroundings. The correct temperature and pH are very important for the proper functioning of enzymes. For example, in your stomach, the environment is highly acidic and there is a specific enzyme called ***pepsinogen***. Pepsinogen will only change into its active form ***pepsin*** when the pH is very low, as it is inside the stomach. Pepsin helps the body break up protein molecules that pepsinogen cannot.

Temperature can also greatly affect the actions of enzymes. In humans, most enzymes work best at the normal body temperature of 37° C, or 98.6° F. When the body temperature increases, these enzymes do not work as well. If the body temperature increases to over 104° F, many of the enzymes will ***denature***, which means they will be unable to function. These examples illustrate how important the correct environment is for enzymes and how specific the actions of enzymes are within an organism.

How Enzymes Work

For all chemical reactions to occur there must be collisions between molecules. Chemical reactions proceed at a faster rate if there are more collisions. Normally the speed of the chemical reaction can be affected by heating the substance, which speeds up molecular motion and, thus, collisions. Increasing the number of molecules in any one area or increasing the concentration of a chemical will also speed up a chemical reaction. In living organisms, increasing the heat and/or increasing the concentration of a particular chemical is usually not an option. This is why enzymes are so important. Enzymes help these chemical reactions proceed faster without doing either one of those options. Enzymes lower what is called the ***activation energy*** of a reaction. The activation energy is the energy needed to get a chemical reaction started. One example of this would be the chemical reaction of a burning piece of paper. If enough heat, possibly from the oxidation chemical reaction of a match being struck, is applied to the paper, the paper will at that point continue to burn at a steady rate as long as there is a supply of burnable material. The paper by itself, though, is very unlikely to start burning without the application of that initial burst of activation energy. How enzymes affect the activation energy of a chemical reaction can also be compared to a sled being pulled up a hill before the ride down the hill. The energy a person needs to pull a sled up a hill would be the activation energy, while the sled going down the hill would be like the occurrence of the chemical reaction. If a friend helped that person drag the sled up the hill, it would take that person less time and energy to get to the top. The friend would be like the enzyme in this comparison. The person could still get the sled to the top of the hill without the friend, but that person would need to use more energy and would not be able to do this as fast on their own.

The outside surface of an enzyme is very important in how it functions and determines what chemicals it can or cannot react with inside the organism. The unique shape of an enzyme allows it to fit closely, almost like two pieces of a jigsaw puzzle, with the chemicals it will react with, which are called the ***substrates***. The substrate molecule chemically reacts with the enzyme at a particular location called the ***active site***. When the substrate molecule connects to this active site of the enzyme, the amount of activation energy needed to start the chemical reaction is lowered. Many times these enzymes are breaking apart or putting together two or more molecules. The biggest factor to consider when you are looking at the functions of enzymes in the body is how greatly they speed up chemical reactions.

Example

The enzyme *catalase* can break up the compound hydrogen peroxide into water and oxygen.

$$2H_2O_2 \rightarrow 2H_2O + O_2$$

In fact, one molecule of the enzyme catalase can break about 40 million molecules of hydrogen peroxide each second.

••••••

Example

Inside the human body you will find the enzyme *carbonic anhydrase*. In order to remove carbon dioxide from the blood it must first be converted to carbonic acid. The chemical reaction that converts carbon dioxide to carbonic acid is very slow; only 200 molecules are converted in one hour.

$$CO_2 + H_2O \rightarrow H_2CO_3$$

But the enzyme carbonic anhydrase increases this rate to 600,000 molecules per second. Without this enzyme your blood would quickly become poisoned and you would die. This means that for this particular example, the enzyme activated chemical reaction is almost 11 million times faster than the reaction without the aid of enzymes!

••••••

Some people are said to be lactose intolerant. These people are not able to digest the sugar in milk called lactose. The intestinal walls should produce an enzyme called ***lactase***, but it does not in people who are intolerant. So, when someone who is lactose intolerant drinks milk or dairy products containing lactose, this sugar that cannot be digested causes the small intestines to spasm. If a lactose intolerant person takes a small amount of lactase prior to drinking milk, they will experience no problems.

Naming Enzymes

In 1876, a German scientist named Wilhelm Kühne found a substance in pancreatic juices that he called *trypsin* that breaks down proteins. He was one of several early scientists who began the long and important study of enzymes.

Most enzyme names are linked to the substrate acted upon. The normal ending of the substrate word is replaced with the ending "ase." For example, most lipids are acted upon by the enzyme lipase. The exceptions to this rule include some of the first enzymes discovered and named by scientists such as Kühne. They include ***trypsin***, ***rennin***, and ***pepsin***. Table 2.5 shows some common enzymes.

Name of Enzyme	Function
Protease	Breaks proteins into amino acids
Lipase	Breaks lipids into fatty acids and glycerol
Cellulase	Breaks sellulose into simpler sugars
Pectinase	Breaks up pectin and other similar plant carbohydrates
Amylase	Breaks starch into simpler sugars

Table 2.5

Practice Section 2-4

1. What is the function of enzymes?
   a) They provide energy for chemical reactions.
   b) The decrease the amount of activation energy needed to begin a chemical reaction.
   c) They transport food in the blood to all cells.
   d) They are chemically destroyed as they break apart molecules within the cell.
2. Which of the following factors affect the rate of enzyme action?
   a) Ph
   c) the amount of enzyme present
   b) temperature
   d) all of these
3. Which of the following is a type of enzyme?
   a) protease b) lipase c) amylase d) all are enzymes
4. What material does the enzyme lactase act upon?
   a) lipids b) proteins c) lactose d) amino acids
5. What is the name of the material that is acted upon by the enzyme?
   a) coenzyme b) inhibitor c) substrate d) fibrin
6. All enzymes are composed of
   a) lipid molecules
   c) protein molecules
   b) sugar molecules
   d) fatty acids
7. What are catalysts?

8. What is the structure of an enzyme?
9. What is the meaning of the "lock and key" idea of an enzyme?

Terms From Chapter 2		
acid	electronegativity	nonpolar covalent bond
activation energy	element	nucleic acid
atom	enzymes	nucleus
atomic number	hydrogen bond	organic compound
base	hydrolysis	periodic table
carbohydrate	inorganic compound	pH scale
chemical bond	ion	polar covalent bond
compound	ionic bond	protein
covalent bond	lipid	proton
dehydration synthesis	molecular formula	structural formula
electron	monosaccharide	valence shell
	neutron	

Chapter 2: Exam

Matching Column for Subatomic Particles

Match the characteristics of subatomic particles with the type of subatomic particle. You may use a choice more than once or not at all.

Characteristics

1. A particle with a positive charge.
2. Determines the atomic number of an atom.
3. Is found outside the nucleus of an atom.
4. Has the smallest mass in the atom.
5. Determines the atomic mass of the atom.
6. Determines the charge of the atom.
7. Is larger in size than an animal cell.
8. A particle with a neutral charge.

Subatomic Particles

a. electron
b. neutron
c. proton
d. protons and neutrons
e. protons and electrons
f. none of these choices
g. protons, neutrons, and electrons

Matching Column for Subatomic Particles

Match the characteristics of the organic molecule with the type of organic molecule that best fits the description. You may use a choice more than once or not at all.

Characterisitcs

9. Is composed of simple sugars.
10. Contains the elements carbon and hydrogen in a ratio of 2:1.
11. Formed from peptide bonds.
12. DNA and RNA are examples.
13. Cellulose is an example.
14. Formed from glycerol and fatty acid molecules.
15. Is also called wax or fat.

Organic Molecule

a. lipid
b. carbohydrate
c. protein
d. nucleic acid

Multiple Choice

16. What elements are found in all organic compounds?
    a) iron and magnesium
    b) carbon and hydrogen
    c) nitrogen and phosphorus
    d) sulfur and carbon
17. Atoms that have the same atomic number but different atomic masses are called
    a) ions
    b) isotopes
    c) isomers
    d) acids
18. H_2O is an example of a
    a) molecular formula
    b) structural formula
    c) both
    d) neither
19. Which group of organic macromolecules is the building block for enzymes?
    a) proteins
    b) lipids
    c) carbohydrates
    d) nucleic acids
20. The formation of a larger, more complex molecule from two smaller molecules, with the removal of water during the process is called
    a) hydrolysis
    b) neutralization
    c) dehydration synthesis
    d) digestion

21. What is the maximum number of electrons that can fit in the second shell?

    a) 2 b) 4 c) 6 d) 8

22. Glucose is an example of a

    a) monosaccharide
    b) disaccharide
    c) polysaccharide
    d) starch

Short Response

23. Describe the structure of a typical atom. What are the subatomic particles, and how are they arranged in an atom?

24. Why do atoms bond together to form molecules? What are the differences between covalent and ionic bonding?

25. Identify each of the following as an element, compound, or a mixture: hydrogen, water, salt, air, carbon.

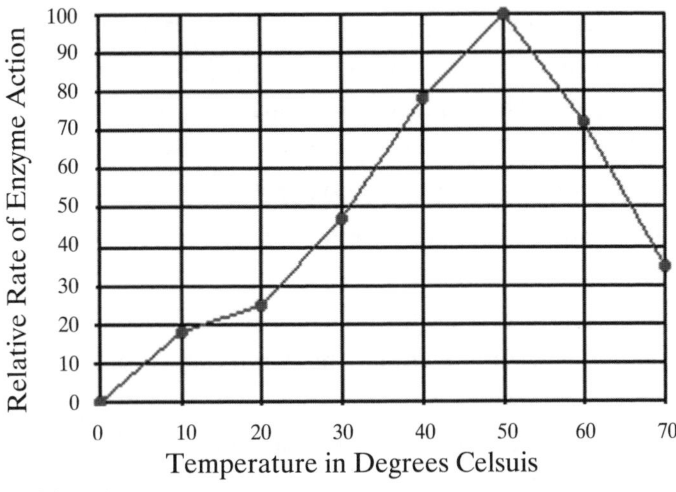

Table 2.6

26. Table 2.6 shows the affects temperature has on a certain enzyme. Study the chart and then answer the questions that follow.

    a) What is the relative rate of enzyme action at 40° C?

    b) What is the temperature at which maximum enzyme action occurs?

    c) What effect do you think raising the temperature above 70° C will have on the rate of enzyme action?

Answer Key

Answers Explained Section 2-1

1. **A** is the correct answer because a proton has a positive charge and is located in the nucleus of the atom.

   B is not correct because a neutron has a neutral charge.

   C is not correct because an electron has a negative charge and is not located in the nucleus of the atom.

   D is not correct because an isotope is the name given to two atoms that have the same atomic number by different atomic masses.

2. **B** is the correct answer. There are six protons found inside the nucleus of an atom with an atomic number of six because the atomic number is equal to the proton number in an atom.

   A, **C**, and **D** are not correct because they are not equal to the proton and atomic number in the atom, which is six.

3. **B** is the correct answer because the charge on the nucleus of an atom is determined by the number of protons found in the nucleus of that atom.

   A is not correct because zero is the charge on the entire atom because three protons at plus one each, and three electrons at negative one each, and zero charge on each of the three neutrons gives you a total net charge of zero.

   C would only be correct if there were six protons in the nucleus of the atom.

   D would only be correct if there were nine protons in the nucleus of the atom.

4. **B** is the correct answer because three neutrons have the most mass, each neutron is equal to about one atomic mass unit (amu) for a total of three amu.

   A is not correct because two protons have less mass; each proton has a mass of about one atomic mass units, so two protons have a mass of two amu.

   C is not correct because 1,000 electrons have less mass; each electron has a mass of about 1/1827th atomic mass units. This means 1,000 electrons have a mass of less than one amu.

5. **B** is the correct answer because the atom is the only particle listed that retains the properties of the element. A, C, and D are not correct because protons, neutrons, and electrons are subatomic particles that do not retain the properties of the element which they are constructed.

6. **C** is the correct answer because the symbol C represents the element carbon. A is not correct because Ca is the symbol for calcium. B and D are not correct because they are not the correct symbol for any elements.

7. The number of protons determines the identity of the atom. Once you know the number of protons in a particular atom, you can identify which atom it is from the Periodic Table.
8. The subatomic particles found in the nucleus of an atom are protons and neutrons. Because protons have a positive charge and neutrons have no charge, the net charge on the nucleus of an atom is positive.
9. The negatively charged electrons are held close to the nucleus of the atom by the positively charged nucleus. Remember positive and negative charges attract one another.

Answers Explained Section 2-2

1. **D** is the correct answer because the second energy shell found around an atom can hold a maximum of eight electrons.

   A, **B**, **C** are not correct because the second shell can hold two, four, or six electrons, but its maximum is eight.

2. **C** is the correct answer because a covalent bond is created by a pair of shared electrons.

   A is not correct because electrons are usually lost by one atom during ionic bonding.

   B is not correct because electrons are usually gained by one atom during ionic bonding.

   D is not correct because **A** and **B** would be a correct answer for an ionic bond.

3. **D** is correct because it is the strongest base present. The pH scale goes from 1 to 14, with 1 being the strongest acid, 7 being a neutral point, and 14 being the strongest base.

   A is not correct because 4 is an acid.

   B is not correct because 8 is a weaker base than 14.

   C is not correct because 12 is a weaker base than 14.

4. **C** is the correct answer because water is a compound, not an element.

   A, **B**, and **D** are not correct because carbon, oxygen, and nitrogen are elements.

5. **B** is the correct answer, because this is the only subatomic particle found outside of the nucleus of the atom.

   A and **C** are not correct because protons and neutrons are found inside the nucleus of the atom.

6. **B** is the correct answer, because adhesion is an attractive force holding two unlike molecules together. In this particular example, a water molecule and a plastic molecule from the shower curtain are being held together.

   A is not correct because cohesion is an attractive force between two like molecules.

   C is not correct because ionic bonding is the attractive force between two ions of opposite charges.

   D is not correct because dehydration synthesis is the joining of two molecules into a larger more complex molecule and the removal of a water.

7. A chemical bond is an attraction between atoms or molecules and allows the formation of chemical compounds, which contain two or more atoms.

8. Ions are charged particles. Positive ions form when atoms lose valence electrons, because now the number of protons is higher than the number of electrons. Negative ions form when atoms gain valence electrons, because now the number of protons is lower than the number of electrons.

9. Water is a polar molecule. A molecule of water is made of two atoms of hydrogen covalently bonded to one atom of oxygen. The electrons are not shared equally between the bonds connecting the hydrogen atoms to the oxygen atoms; this causes the two hydrogen atoms to be slightly positive in charge and the oxygen slightly negative. Water becomes a very good liquid solvent because many other molecules necessary for life are also polar in nature (sugar, salt). Polar solutes dissolve very easily into polar solvents.

Answers Explained Section 2-3

1. **B** is the correct answer because part of the enzyme structure is created from protein molecules.

   A, **C**, and **D** are not correct because they are organic molecules but not part of the enzyme structure.

2. **C** is the correct answer because starch is a polysaccharide.

   A and **B** are not correct because glucose and fructose are simple sugars.

   D is not correct because sucrose is a disaccharide.

3. **D** is the correct answer because glucose is not a subunit of nucleic acids. Glucose is a subunit of carbohydrates and the subunit of nucleic acids are nucleotides.

   A, **B**, and **C** are not correct because they are subunits that are correctly matched with the organic molecules they compose.

4. **B** is correct because carbon is the major element used to identify organic compounds.

A, **C**, and **D** are not correct because nitrogen, iron, and oxygen can be found in organic molecules, but there are many organic molecules that do not contain these elements.

5. **B** is the correct answer because dehydration synthesis is the combining of smaller molecules into a larger molecule with the removal of water.

   A is not correct because hydrolysis is the breaking apart of large molecules into smaller molecules with the addition of water.

   C is not correct because metabolism is the sum total of the chemical processes occurring inside of an organism.

   D is not correct because ion formation is when an atom acquires a charge by either losing or gaining an electron.

6. **D** is the correct answer because DNA and RNA are types of nucleic acids.

   A, **B**, and **C** are types of organic molecules but are not structural components of DNA and RNA.

7. Organic chemistry is a discipline within chemistry involving the study of the structure, properties, composition, reactions, and preparation of carbon-based compounds. The four major carbon based compounds important in the study of biology are carbohydrates, proteins, lipids, and nucleic acids.

8. The basic unit of all proteins is the amino acid. A peptide bond is the name of the chemical bond that connects amino acid molecules together to form protein molecules.

9. Nucleic acids are made up of a subunit called nucleotides. Nucleotides are chemically constructed of one molecule of sugar (DNA or RNA), bound to one molecule of phosphate, and to one nitrogen-containing base (adenine, uracil, cytosine, guanine, or thymine).

Answers Explained Section 2-4

1. **B** is the correct answer because enzymes decrease the amount of activation energy needed to begin a chemical reaction.

   A is not correct because enzymes do not provide energy for chemical reactions.

   C is not correct because enzymes do not transport food in the blood to cells; this job is achieved by the blood plasma.

   D is not correct because enzymes do break apart molecules in the cell but are not destroyed in the process.

2. **D** is the correct answer because **A**, **B**, and **C** (pH, temperature, and the amount of enzyme present) affect the rate of enzyme action.

3. **D** is the correct answer because **A**, **B**, and **C** (protease, lipase, and amylase) are examples of types of enzymes.
4. **C** is the correct answer because lactase acts on the substance lactose.

   A is not correct because lipids are acted upon by a class of enzymes called lipases.

   B is not correct because proteins are acted upon by a class of enzymes called proteases.

   D is not correct because amino acids are the building blocks of proteins.
5. **C** is the correct answer because enzymes act on materials called substrates.

   A is not correct because coenzymes are chemicals, usually vitamins, that aid the action of enzymes.

   B is not correct because inhibitors are chemicals that slow or block a chemical reaction.

   D is not correct because fibrin is the chemical released by the platelets in the blood that begins the clotting process.
6. **C** is the correct answer because enzymes are composed of protein molecules.

   A, **B**, and **D** are not correct because they are all organic molecules but not the structural components of enzymes.
7. Catalysts are chemical substances that reduce the activation energy of a chemical reaction. This action increases the speed of the chemical reaction. The catalyst is not consumed by the reaction.
8. Enzymes are protein based molecules (long strands of chemically connected amino acids) whose activities are determined by their three-dimensional structure.
9. All enzymes operate under the "lock and key" idea. Because of their unique three-dimensional structures are very specific to the chemical reaction they can influence (connect with structurally). Just like a lock has only one key that will make it work properly, most chemical reactions have only one type of enzyme that will allow the reaction to work at a faster rate.

Answers Explained Chapter 2 Exam

Matching Column

1. C	4. A	7. F	10. B	13. B
2. C	5. D	8. B	11. C	14. A
3. A	6. E	9. B	12. D	15. A

Multiple Choice

16. **B** is correct because only the elements carbon and hydrogen would be found in all compounds considered to be organic.

    A, **C**, and **D** are not correct because even though some of these elements are found in some organic molecules, there are many organic molecules that do not have both of these elements in them.

17. **B** is correct because two atoms that have the same atomic number (proton number) but have different atomic masses would be considered isotopes of one another.

    A is not correct because an ion is an atom that has a charge.

    C is not correct because an isomer is two compounds that have the same molecular formula but different structural formulas.

    D is not correct because an acid is any chemical that in a liquid solution will give off H^+ ions.

18. **A** is correct because a molecular formula shows the number of each element present in a compound.

    B is not correct because a structural formula shows how each element in a compound is bonded to every other element in the compound.

19. **A** is correct because the structural units of all enzymes are proteins.

    B is not correct but lipids are important structural units of cell membranes.

    C is not correct but carbohydrates are important energy sources for the cell.

    D is not correct but nucleic acids are used to create genetic material.

20. **C** is correct because the removal of water to join two or more compounds together is the definition of dehydration synthesis.

    A is not correct because hydrolysis is the opposite of dehydration synthesis, where water is added to a larger compound to split it up into smaller compounds.

    B is not correct because a neutralization reaction is when you combine an acid and a base to produce water and a salt compound.

    D is not correct because digestion is an example of hydrolysis.

21. **D** is the correct answer because the second electron shell can hold at most 8 electrons.

    A, **B**, and **C** are not correct because of what was said for answer above.

22. **A** is correct because glucose is a monosaccharide, which is a single sugar compound (simple sugar).

    B is not correct because disaccharide means double-sugar.

C is not correct because polysaccharide means multiple sugars bonded together.

D is not correct because starch is an example of a polysaccharide.

Short Response

23. The structure of a typical atom would have the protons and neutrons in the nucleus or center of the atom, and the electrons spinning in a cloud arrangement outside of the nucleus of the atom.

24. Most atoms bond to other atoms to become more stable. If an atom does not have a full valence electron shell, it is unstable. These unstable atoms will seek electrons from other atoms to complete their valence shell. Sometimes the bonds formed are covalent (the two unstable atoms will share the electrons to become stable) and other times ionic (one atom takes the electron the other releases the electron to make them both stable), it depends on the atoms in question.

25. **Hydrogen** is an example of an element.

    Water is an example of a compound.

    Salt is an example of a compound.

    Air is an example of a mixture.

    Carbon is an example of an element.

26. a) about 78%

    b) 50° C

    c) As the temperature continues to get warmer, the rate of enzyme action will continue to decrease until the enzymes denature.

3

The Cell

In this chapter you will learn an important theory that contains the idea that the basic unit of ***structure*** for all life is the cell. This means that all life on this planet is made up of cells, no matter how small or how large that organism may be. Another idea that is part of this cell theory is that the cell is the basic unit of ***function*** for all life. This means that all living things do what they do because their actions are important to keep the cells that make up their body alive. Let us start by taking a glimpse at how the cell was first discovered.

Lesson 3-1: The Discovery of the Cell

The study and discovery of the cell are closely linked with advances in scientific equipment. One of the most important pieces of equipment that led to much of our current knowledge about cells is the microscope. Because of the small size of most cells, the use of the microscope is critical if they are to be accurately observed.

History of the Microscope

The earliest development of the microscope took place a long time after people had discovered the magnifying power of lenses. A container of water, a bead of glass, a gemstone can all magnify objects to some degree and can be considered early lenses.

Roger Bacon (1214–1294) is given credit for introducing the magnifying lens as eye glasses, or spectacles, as they were known during his time, in 1268. This is probably the first recorded time a magnifying glass was used to support human eyesight. It would take several hundred years before these glass lenses were used to make distant objects appear closer or small objects appear bigger. The invention of the telescope was in 1608 by a Dutch lens maker named Hans Lippershey. The microscope was probably created a few years earlier, around 1597, by Zaccharias Jansseen and his son Han, who were also Dutch lens makers. This father-and-son combo created the earliest documented compound microscope by focusing light through two lenses. Robert Hooke would later refine the structure of the compound microscope.

Anton van Leeuwenhoek (1632–1723), a naturalist and lens maker, constructed a very powerful lens to view small objects. Van Leeuwenhoek created large single lens microscopes that could magnify objects up to 270X their normal size. With these lenses, he saw small objects that had never been seen by human eyes before, such as bacterial cells, protozoa, and blood and yeast cells.

Robert Hooke (1635–1703), a physicist by trade, refined the structure of the compound microscope. As mentioned earlier, a compound microscope is a microscope where the object is now viewed through two lenses, instead of just one. Robert Hooke is given credit for first coining the word "cell," when he described the many small, interconnected boxes (which looked like little prison "cells") he saw under the microscope while looking at a piece of cork. Cork is made from tree bark, so it consists of dead plant cells. Hooke, although not known to him at the time, was viewing these dead plant cells.

In 1831, Robert Brown, a botanist, was the first to describe the dark, circular object found at the center of all plant cells as a **nucleus**.

In 1838, Matthias Schleiden, a botanist, and Theodor Schwann, a zoologist, researched that all living things are made of cells, and in 1855, Rudolph Virchow added that all cells arise only from other living cells. The work of these three scientists led to what is today known as the ***cell theory.***

Summary of the Cell Theory

1. The cell is the unit of structure for all living things.
2. The cell is the unit of function for all living things.
3. All cells arise only from other living cells.

••••••

This theory is used as the basis for defining life. Cells perform the chemical life-processes necessary for life. Anything that is considered by scientists to be alive must be constructed of cells. And finally, the only way a living organism can reproduce is by the actions of the cells within its body.

The cell theory contains the guiding principles for the study of life. But there are some things present in the study of biology that do not meet the principles of this theory.

Some Exceptions to the Cell Theory

1. According to the cell theory, all living things are made of cells. A virus is not made of cells. This means that either a virus is a living thing that does not follow this principle of life being made of cells, or viruses are not living. Viruses are made of two chemicals: protein and nucleic acids. They appear to reproduce, but only when they are inside of living cells.
2. According to the cell theory, all cells come from other cells. This means that only cells have the ability to replicate new life. But there are two organelles found in cells that contain their own genetic information and can replicate on their own: chloroplasts and mitochondria. One theory is that chloroplasts and mitochondria were, at one time, independent types of cells that developed a close interdependence with the ancient form of the modern-day eukaryotic cells, and eventually these two cells became one new form of cell: the present-day eukaryotic cell.
3. According to the cell theory, all cells come from other cells. So, the question unanswered is: "How did the first cell get on Earth?" The answer to this question would be at the essence to the origins of life on this planet. This will be discussed more in Chapter 5.

••••••

Practice Section 3-1

1. Who was the first to use the word "cell"?
   a) Van Leeuwenhoek c) Brown
   b) Hooke d) Schwann
2. Which of the following statements is NOT part of the cell theory?
   a) All living things are made of cells.
   b) Cells are the unit of function for all life.
   c) All cells contain a nucleus.
   d) All cells come from other cells.
3. The invention of what scientific instrument led most directly to the formation of the cell theory?
   a) ultracentrifuge b) balance c) microscope d) chromatography
4. Which of the following organelles contain its own DNA.
   a) cell membrane b) vacuole c) ribosome d) mitochondria

5. All of the following are made of cells except
   a) plants b) bacteria c) viruses d) animals
6. Which of the following scientists' work was not involved in the formation of the cell theory?
   a) Schwann b) Darwin c) Virchow d) Schleiden
7. What is the cell theory?
8. Why are viruses considered by some scientists to be an exception to ideas held in the cell theory of biology?
9. What was the historical importance of Robert Hooke's observation of a slice of cork under a microscope?

Lesson 3-2: The Cell Is the Unit of Structure and Function

All living things are made of cells. Cells are the structural units of all life. Some living things are made up of only one cell (bacteria and some protists), and some living things are made up of many cells working together (humans, fish, and dogs). There are two major types of cells: **prokaryotes** and **eukaryotes**.

Comparisons Between Prokaryotes and Eukaryotes

Prokaryotes	Eukaryotes
No nucleus that surrounds the genetic information. KNA not coated with protein.	Membrane-bounded nucleus that contains genetic information in the form of chromosomes.
Cell division is accomplished by binary fission or budding.	Cell division is accomplished by mitosis.
No tissue development of these cells.	Some cells of this variety have extensive tissue development.
Cells of this type are anaerobic and aerobic.	Most cells of this type are aerobic.
Metabolism is achieved in many different forms.	Nutrition is heterotrophic or autotrophic.
Cell size (1–10 μm)	Cell size (1–10 μm)
Mitochondria absent. Respiration at times occurs in the cell membrane.	Mitochondria present. Aerobic respiration occurs in the mitochondria.
Ribosomes present in the cytoplasm.	Ribosomes are present in the cytoplasm and on the endoplasmic reticulum.

Table 3.1

Eukaryotes are cells that have a nucleus and many membrane-bound organelles. The presence of these membrane-bound organelles allows the eukaryotic cell to compartmentalize chemical reaction within the cell structure.

Prokaryotes are cells that do not have a nucleus to surround their genetic information and do not have membrane-bound organelles. Many prokaryotic cells have a cell wall made of ***peptidoglycan***. Although prokaryotic cells are simpler in structure than eukaryotic cells they have shown an ability to adapt very quickly to changing environments on the Earth for the past 3 billion years.

Table 3.1 on page 72 summarizes some of the major comparisons between these two types of cells.

Although there are many different types of cells that have many different functions, all cells have structural similarities. It is important to know something of these small structures, or **organelles**, found within most cells to best understand how cells live.

Parts of the Cell

A thin organic layer called plasma, or **cell membrane**, surrounds all cells. This membrane defines the size and, in some cases, the shape of the cell. The cell membrane, which has very tiny channels going through its surface, controls the flow of materials into and out of the cell. The cell membrane is composed of a double lipid layer with large protein molecules inside of it. Some of the protein molecules act as channels by which small particles can flow through.

The cell membrane of a cell is said to be ***selectively permeable***. Selectively permeable means that some materials can pass through it easily, others can not. Usually very small molecules, such as amino acids, glycerols, and fatty acids can pass through the cell membrane while larger molecules, such as proteins and starches, can not pass through. The charge of the molecule in some cases determines if it can pass through the membrane, with nonpolar substances being more easily let into the cell than polar molecules.

When you look at most cells under a compound microscope, inside of these cells you will see a dark small circle. This dark circular object is the ***nucleus*** of the cell. This structure controls all the chemical activities of the cell as well as the reproduction of new cells. The nucleic acid material called DNA and RNA are the principal molecules responsible for these actions.

A jelly-like, clear substance fills the entire cell's volume and is called **cytoplasm**. Cells are about 90% cytoplasm. The cytoplasm contains proteins, amino acids, and glucose molecules, but it mostly contains water. Cytoplasm is where many of the chemical reactions inside the cell take place and where the rest of the organelles reside.

Cell Organelles Found Inside the Cell and Their Functions

Nucleus	Organelle found in plant and animal cells that contains the genetic information of the cell and controls chemical activities that take place inside the cell.
Cell membrane	A structure consisting of proteins and lipids that surrounds all cells and controls the flow of materials into and out of the cell.
Cell wall	A supportive structure found surrounding the cell membrane of plant, bacterial, and fungi cells. Materials pass easily through this organelle.
Cytoplasm	A gel-like substance composed of about 90% water and dissolved materials such as sugars, amino acids, and salts, that provides a watery environment for chemical reactions to occur in the cell.
Endoplasmic reticulum	A collection of membranes that runs throughout the cytoplasm of eukaryotic cells that compartmentalizes sections of this cell for specific chemical reactions to occur. The walls of some E.R. also contain ribosomes.
Golgi bodies	An organelle made up of stacks of flattened memebranes that package chemicals for removal from the cell.
Ribosomes	Smallest organelles found in the dell and are the site of protein synthesis. These organelles can be found attached to the E.R. or floating in the cytoplasm. Eukaryoteic and prokaryotic cells contain these organelles.
Vacuoles	An organelle that stores food in plants and animals. Contractile vacuoles are also used in some one-celled organisms as a means of water regulation within the cell.
Lysosomes	An organelle found in plant and animal cells that contain digestive enzymes.
Mitochondria	The organelle found in plant and animal cells where aerobic respiration and energy production occurs.
Chloroplasts	This organelle is found in photosynthetic plant cells and is the site of food production.
Microtubules	Thin cylinder-shaped structures of various sizes found inside many cells and supports the shape of certain cells like a skeleton.

Table 3.2

The **endoplasmic reticulum** is a series of channels that are found throughout the cytoplasm of the cell. The surfaces of these channels provide space for important chemical reactions and these channels help funnel chemicals to proper destinations inside the cell.

Ribosomes are the smallest organelles and are found either embedded in the endoplasmic reticulum or floating freely in the cytoplasm. Most ribosomes function in the creation of proteins. Many of these proteins are then used to create parts of the cell.

The **mitochondria** are oval-shaped structures located inside the cell that contain enzymes that aid the cell in drawing energy from food. The energy released from food is stored in a chemical called ATP.

There are bag-like structures located in the cytoplasm of cells called **vacuoles** that store water and other materials for use by the cell. In plant cells, these vacuoles are very large; in animal cells, they are much smaller in size.

Lysosomes are small saclike structures that contain digestive enzymes, which are used to dissolve large food molecules, and to break up old or damaged cell structures. The lysosomes are also responsible for the digestive character of certain white blood cells and the destruction of tissue in some animals such as the tail of a tadpole.

The **Golgi bodies** are a group of organelles that prepare certain protein compounds for dispersal from the cell. These protein chemicals are enclosed in small membranes for this journey.

Under very powerful microscopes, small, thin support tubes can be seen crossing the inside of the cell from membrane to membrane. These tubes are called **microtubules** and they act as support structures for the cell. These microtubules will sometimes extend beyond the outer cell membrane to the outside of the cell. Many times these tube extensions help propel the cell from place to place, almost like the fins of a fish. **Cilia** and **flagella** are two specific microtubule extensions found on some cells.

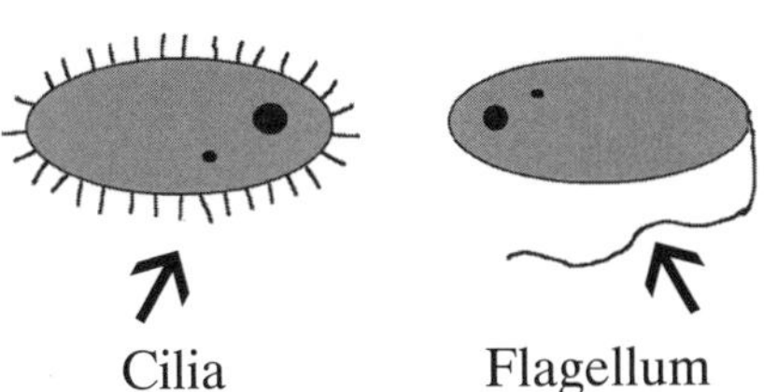

There are some organelles that you will only find commonly in certain types of cells. For instance, all plant cells are not only surrounded by a cell membrane, but also a more solid structure framing the cell membrane called a **cell wall**. The cell walls of plant cells are composed mainly of the complex carbohydrate cellulose. The cell wall acts as a supportive structure for plant cells. Plant cells also contain ***chloroplast*** organelles. Chloroplast organelles contain the pigment **chlorophyll** and are the site of the chemical process of photosynthesis.

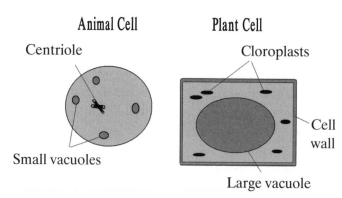

Animal Cell

Centriole

Small vacuoles

Plant Cell

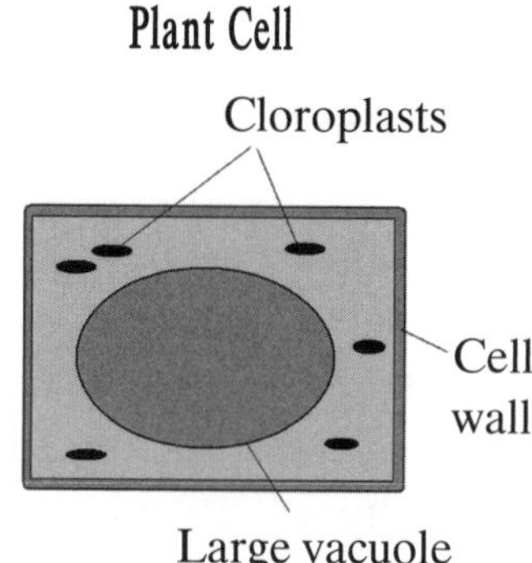

In animal cells only, you will find an organelle called a **centrosome**. The centrosome contains two smaller structures called **centrioles**. These centrioles are important for the division of DNA during cell reproduction.

Practice Section 3-2

1. Which of the following statements is true about prokaryotic cells?
   a) All prokaryotic cells grow and reproduce.
   b) All prokaryotic cells have cell membranes and cytoplasm.
   c) All prokaryotic cells lack a nucleus.
   d) All of the above are true of prokaryotic cells.
2. In what type of cell do you NOT find a cell wall?
   a) bacterial cells b) plant cells c) animal cells d) fungi cells
3. What is the name of the saclike cell organelle responsible for storing materials?
   a) cell membrane c) mitochondria
   b) endoplasmic reticulum d) vacuole
4. What chemical process occurs in the cell organelle called the chloroplast?
   a) respiration b) excretion c) photosynthesis d) chemosynthesis
5. Which structure regulates the movement of materials into and out of the animal cell?
   a) cell wall b) cell membrane c) nucleus d) ribosome
6. What statement about ribosomes is true?
   a) Ribosomes contain DNA.
   b) Ribosomes are the largest organelle in the cell.
   c) Ribosomes are the site of respiration.
   d) Ribosome are the site of protein synthesis.
7. What is the main function of the plasma or "cell membrane"?
8. What are the locations in the cell where the organelles called ribosomes can be located?
9. Which organelle in the cell performs the action of intracellular digestion?

Lesson 3-3: Transport and the Cell

The cell is the basic unit of structure and function for all living organisms. This means the cell is constantly performing chemical reactions important for life. Most of these chemical reactions require that materials such as water, oxygen, and food be imported into the cell. The reactions that take place with these chemicals result in products such as carbon dioxide, excess water, and nitrogenous wastes that must be removed from the cell. This means that the proper flow of materials into and out of the cell is critical in maintaining homeostasis and the health of the cell.

Structure of the Cell Membrane

All substances that go into or come out of the cell must in some way react with the cell membrane. The movement of materials across this membrane is a form of transport. In order to best understand how transport of materials occurs across this membrane, let's take a closer look at its structure.

Structure of the Cell Membrane

Outside the cell
Phospholipids
Tail "nonpolar"
Head "polar"
Proteins
Inside the cell
Channel protein

As you can see from the diagram the cell membrane is composed of what is called a ***phospholipid bilayer***. There are two layers of this lipid material that create the walls of the cell membrane. The molecules that make up this double-lipid layer have round head-like structures and two tail-like structures.

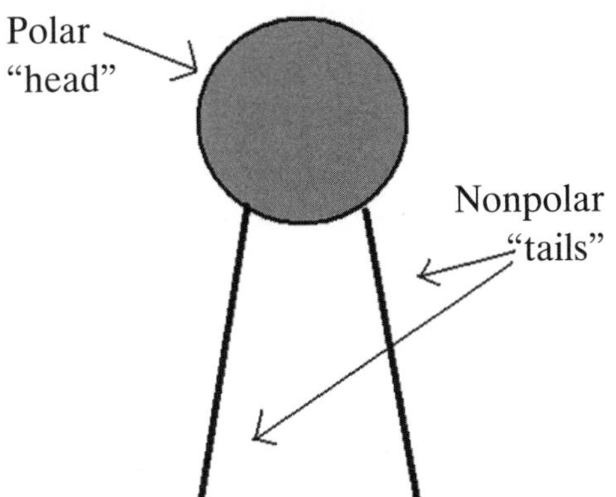

The heads of these phospholipid molecules are ***hydrophilic***, or chemically attracted to water. The tails of these same molecules are ***hydrophobic***, or chemically repelled by water. This causes the molecules within the cell membrane to line up perfectly based on these chemical properties. The heads face the watery environment inside and outside of the cell, while the tails face the area between the heads of the phospholipid molecules that is free of water. This condition forms the boundary for the cell internal environment.

Protein molecules are scattered throughout the cell membrane and embedded into the phospholipid bilayer. Some of these protein molecules are partially embedded into the lipid membrane with only their upper sections exposed on the outer membrane. These partially exposed protein molecules usually have the function of cell identification. Along with carbohydrate molecules on the lipid surface, they help cells in the body recognize one another. Some protein molecules embedded in the membrane are long enough to have their tips exposed on each side of the lipid bilayer. These longer proteins are used to channel certain molecules into and out of the cell, which are not able to work their way through the double lipid section of the membrane.

This model of the cell membrane is sometimes called the ***fluid-mosaic model.*** It is called fluid because the membrane is flexible and allows certain materials to pass through it. The cell membrane is called a mosaic because of the many protein molecules dotting the surface of the membrane making it look like the tiles of a mosaic painting.

Types of Transport

All molecules are in constant motion. Molecules are always constantly moving and bumping into other molecules that, in turn, cause the molecules to spread out. **Passive transport**, or **diffusion**, is a movement of materials from an area where the molecules are highly concentrated to an area where they can spread out or are in lower concentrations.

In passive transport, because of the natural movement and collision of molecules, the cell does not have to provide any energy for the materials to disperse within the cell or into the cell. These molecules, on there own, will evenly distribute themselves. For example, suppose a cell with a low concentration of sugar in its cytoplasm is placed into a solution that has a high concentration of sugar. The sugar molecules that are in a high concentration outside of the cell will move by passive transport into the cell where the sugar concentration is

Diffusion of a Substance in Water

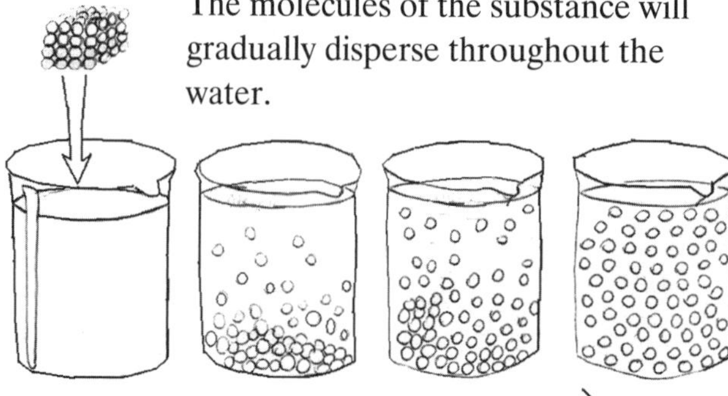
The molecules of the substance will gradually disperse throughout the water.

lower. Passive transport can move certain molecules into the cell across the bilipid layers of the cell membrane. Molecules that can enter across the cell's bilayers must be small and of the correct polarity. Water, lipids, and other hydrophobic substances are some of the materials that enter the cell in this fashion.

Materials such as simple sugars and amino acids can also enter by the cell by passive transport, but because of the polarity of these substances, they cannot cross the lipid bilayers of the cell membrane. This is where a specific type of substance that is part of the cell membrane called a ***channel protein*** aids in transport. These channel proteins are the proteins mentioned earlier that completely span the double lipid membrane. Inside of these proteins are small channels that run from the watery environment outside of the cell to the environment inside the cell. Small materials that need to enter the cell but are water-soluble can enter by way of the channel proteins.

Any time two solutions of varying concentrations are being compared, the solution with the higher concentration is said to be ***hypertonic***, while the solution with the lower concentration is said to be ***hypotonic.*** If two solutions are compared that have equal concentrations of a particular solute, they are said to be ***isotonic*** in relation to each other. When two solutions have the same concentration of dissolved particles, there is no net movement of molecules from one solution to the other. The solutions are said to be in equilibrium, but this does not mean that the molecules are not moving from one solution to the other. In fact, many molecules are still moving between the solutions but the number of molecules moving in each direction is equal.

Osmosis is the diffusion of water. Water is the major constituent of most living things. This means that the movement of water is very important for the maintenance of living systems. If a one-celled organism is in a watery environment that is full of dissolved particles, the concentration of water outside the cell will probably be lower than the concentration of water inside the one-celled organism. This means that osmosis will cause water to flow into the cell from the environment. If too much water enters a one-celled organism, it can burst, so many of these microorganisms have special structures called ***contractile vacuoles*** that pump water out of the cell.

Active transport is the movement of molecules from an area of low concentration to an area of high concentration. These molecules must be directed to an area that normal molecular motion would not move these molecules. In active transport, the energy in the form of ATP is used to corral these molecules into tight quarters, even though their constant movement is working against such an action. Active transport requires that the cell expend energy to accomplish the movement of molecules to a certain area, while passive transport does not. It might be helpful to think of a preschool teacher trying to get her class of 30 students into a straight line after snack time. It takes very little energy from the teacher to let the students run around the playground (passive transport), but it would take much more energy to get them into a specific set position; for example, a line (active transport). On the surface of the cell there are chemicals called ***carrier proteins*** that can bind to materials from the cell's environment and move them across the membrane. This type of transport requires energy and usually moves materials from a hypertonic solution to a hypotonic solution.

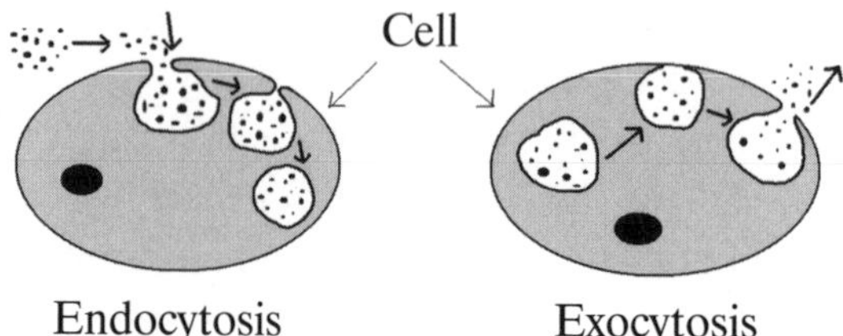

There are some cells that need to move very large particles inside the membrane. A form of active transport called **endocytosis** is when the cell membrane will actually surround the material and form a vacuole that will move within the cell. The materials in this process never actually cross the membrane but are wrapped within it. **Exocytosis** is a form of active transport where large particles are expelled from the inside of a cell in the exact reverse process of endocytosis.

Movement Within the Cell

Substances must also be moved inside of the cell from location to location. The natural diffusion of molecules in many cases is the answer to this problem. The small volume of most cells allows the slow process of diffusion to be sufficient for this purpose. In fact, it is because diffusion is the common form of transport inside of cells that the size of cells must remain small. If a cell with a large volume had to use diffusion to move particles, it would not survive long because of the time it would take to transport these molecules. Most one-celled organisms and the cells that make up multicellular organisms have an average limit to their size because of this very fact.

In larger, more complex, multicellular organisms, such as humans, special transport mechanisms must be created to disperse molecules throughout their bodies. These systems are called ***circulatory systems***, and in many cases involve networks of tubes that aide by pumping-mechanisms forcefully bringing these substances to places far away in the body in much faster rates than diffusion ever could.

Practice Section 3-3

1. Which of the following is true about the cell membrane?

   a) The cell membrane regulates the movement of materials into the cell.

   b) The cell membrane is composed of lipid and protein molecules.

c) The cell membrane surrounds all living cells.

d) A, B, and C are true about the cell membrane.

2. Diffusion is

   a) the movement of molecules from an area of high concentration to an area of low concentration.

   b) the movement of molecules from an area of low concentration to an area of high concentration.

   c) when molecules do not move from their positions.

   d) the movement of food into the cell with the expenditure of energy.

3. A solution that is hypertonic has

   a) less dissolved materials than a hypotonic solution.

   b) more dissolved materials than a hypotonic solution.

   c) the same amount of dissolved materials as a hypotonic solution.

   d) none of the answers are correct.

4. Which of the following processes requires that the cell expends energy?

   a) osmosis c) active transport

   b) diffusion d) passive transport

5. The cell membrane's bilayer is made of

   a) protein molecules. c) carbohydrates.

   b) phospholipids. d) nucleic acids.

6. What is the name of the process of taking larger materials into the cell by means of infoldings of the cell membrane?

   a) endocytosis b) exocytosis c) osmosis d) diffusion

7. What are the chemical substances that make up the cell membrane?

8. What is the major difference between active and passive transport?

9. Why is the process of endocytosis considered to be a form of active transport?

Lesson 3-4: Reproduction and Division of the Cell

The ability of all life to replicate itself for future generations originates in reproduction of the cell. As stated in the cell theory (Lesson 3-1), all cells arise from other cells. This holds true for one-celled organisms replicating their entire body, to multicellular organisms generating more cells within their larger, more complex bodies. In order to best understand how all cells replicate, let's take a look at what is called the ***cell cycle***.

How long a cell lives and the amount of time the cell spends in each different stage can vary from cell to cell, but some general statements can be said of this process. All cells spend most of their lives (about 90%) in what is called **interphase**. Interphase is divided into three stages: G1, S, and G2. In G1, the cell experiences normal growth in volume and carries on normal processes of life. In the S phase, the genetic information (DNA) is doubled, providing the correct amount of this material for equal distribution during cell division. In the G2 phase, the cell is chemically preparing for cell division by replicating organelles and creating chemicals needed for the actual division process.

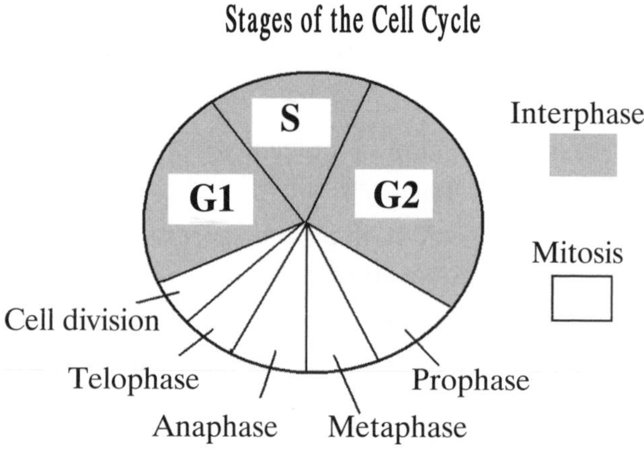

Mitosis

When interphase has ended, the cell begins the process of genetic division and cell division. **Mitosis** is the process by which the genetic material of a cell is equally divided into two complete sets.

The time the cell spends in mitosis can be subdivided into four phases: **prophase, metaphase, anaphase, and telophase.**

In prophase, the genetic information that was doubled during the S stage of interphase begins to coil, thicken, and shorten, in order to be more easily moved during the process of cell division. When the genetic information has coiled into visible rod-shaped structures, which can be seen under the compound microscope, the material is called **chromosomes**. A chromosome at prophase consists of joined ***sister chromatids***, which are the doubled DNA structures from the S stage of interphase. The structure that joins the sister chromatids is called a ***centromere***. Prophase is the stage of mitosis that takes the longest time to complete. Small structures called ***centrioles*** move to each pole of the cell during this phase. Centrioles line up long, thin, wire-like structures called ***spindle fibers*** across the length of the cell from pole to pole. These spindle fibers help line up the chromosomes during the next phase of mitosis called metaphase.

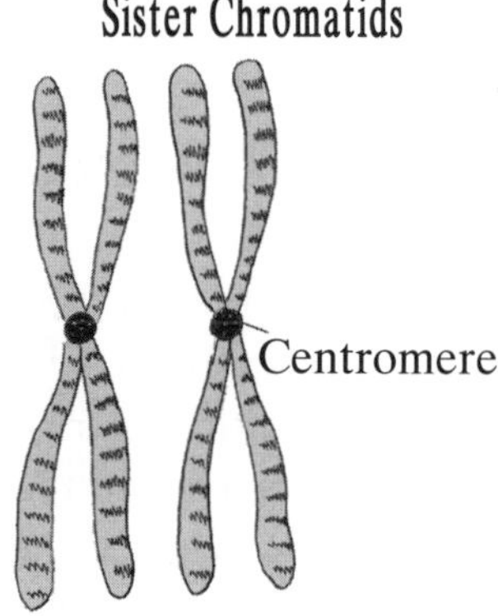

In metaphase, the doubled chromosomes line up at the center of the cell. In anaphase, the doubled chromosomes split and begin to move towards each corner of the cell. In telophase, the split genetic information, now in each corner of the cell, begins to uncoil and lengthen once more and the cell begins to divide into two new cells. ***Cytokinesis*** is the name given to the division of the cytoplasm of one cell into the two new cells. In animal cells, cytokinesis is conducted by the cell membrane pinching in at the equator until it eventually has formed two new cells. In plant cells cytokinesis is performed by a structure called a ***cell plate***, which forms down the center to create two new cells. All one-celled organisms reproduce by mitosis. Mitosis is used by multicelled organisms for the growth and repair of all the cells in their bodies except cells called **gametes**.

Meiosis

Gametes are the sex cells found in the bodies of multicellular sexual organisms. All human cells are produced by mitosis except the sperm and egg, which are the human sex cells, or gametes. Sex cells divide in a special process called **meiosis**. Meiosis is a process of cellular division where the genetic information is reduced in half for the newly created cells. In sexual reproduction genetic information from the male and female are put together to produce a new organism. The new organism must have the same number of chromosomes as each of the parents. Meiosis ensures that each parent donates only half of the needed chromosomes, by aid of the sex cell, to the offspring. When each parent donates half of the needed chromosomes, the final offspring will have the full number of chromosomes after fertilization. All body cells that have a full number of chromosomes are said to be **diploid**. Sex cells have half the number of chromosomes as diploid cells and are called ***haploid*** or **monoploid.**

The actions of interphase, during the cell cycle, are virtually the same for sex cells and body cells. It is when the sex cell begins the process of cell division or meiosis that some significant differences occur compared to the process of mitosis of body cells. Meiosis is subdivided into stages that show the division and reduction of genetic information. In the first stage called prophase I, the cell begins to coil and thicken its doubled genetic information. Then in a step unique to meiosis, each pair of sister chromatids matches up with another set of sister chromatids. This action forms a four-part structure called a ***tetrad***, which are actually two sets of sister chromatids.

Formation of a Tetrad (Sister Chromatids Are Joined)

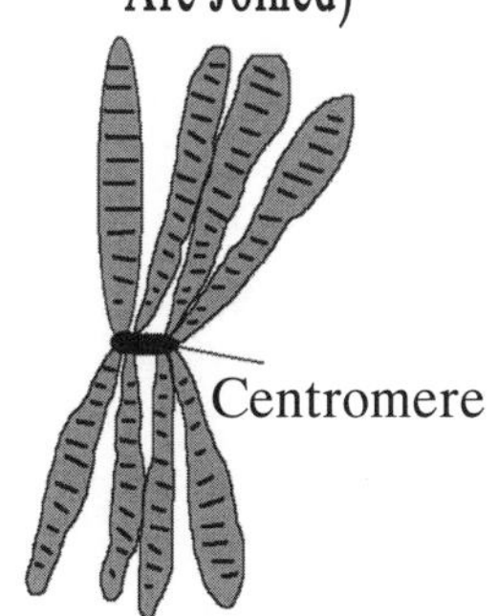

When these sets of sister chromatids join together to form tetrads, some genetic information is exchanged between the chromatids. This exchange of genetic information is called **crossing over**. The process of crossing over makes the chromosomes more genetically diverse, which leads to more genetically diverse offspring.

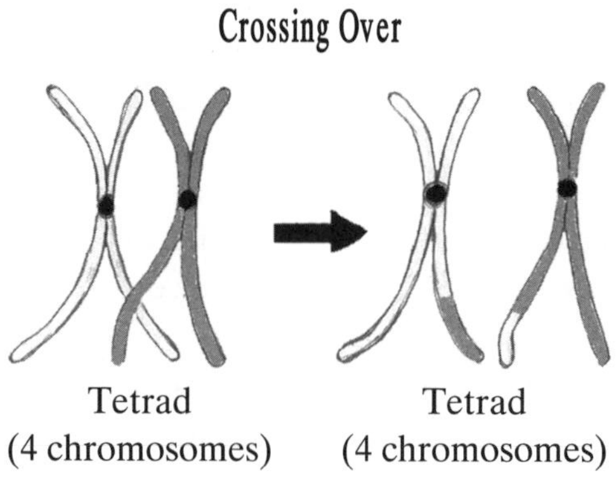

The tetrads line up at the center of the cell in preparation for division during the second stage of meiosis called metaphase I. This is a difference from metaphase in mitosis where just the sister chromatids line up at the center. In the third stage of meiosis, called anaphase I, the tetrads separate into pairs of sister chromatids and move toward the poles of the cell.

In telophase I, the sister chromatids uncoil and the cell begins to break into two new cells. At the end of this phase, there are two new cells with half the number of chromosomes (monoploid) from the original parent cell (diploid). Even though the chromosome number is halved, there is still double the amount of these chromosomes necessary for the final product. This is why the cell undergoes the four stages—prophase, metaphase, anaphase, and telophase—one more time. This action brings the double chromosomes, called sister chromatids, to single chromosomes. The second stages of meiosis are called prophase II, metaphase II, anaphase II, and telophase II. These four stages ensure that each of the four cells created has a single copy of each chromosome.

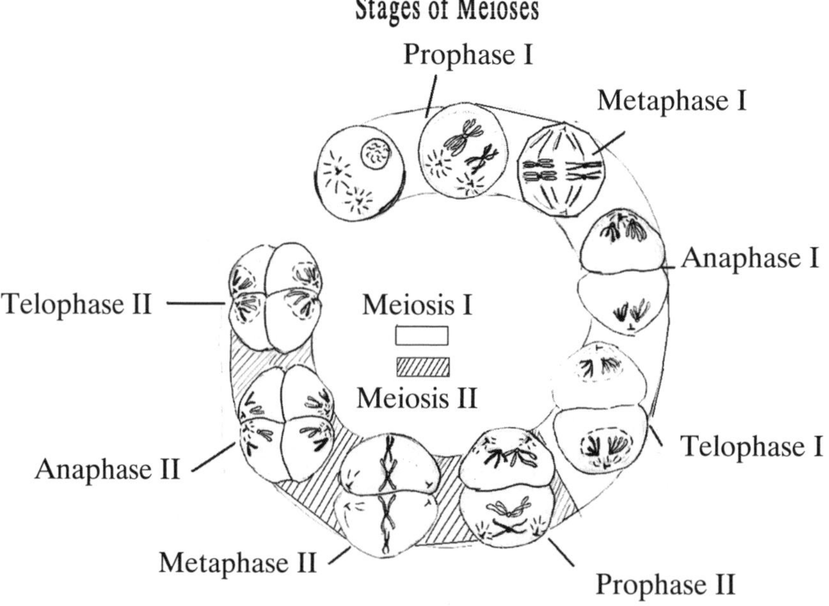

Spermatogenesis and Oogenesis

The process in human males where the sex cells divide and mature into haploid sperm cells is called **spermatogenesis**. A human male will produce millions of sperm in this fashion from puberty almost until death.

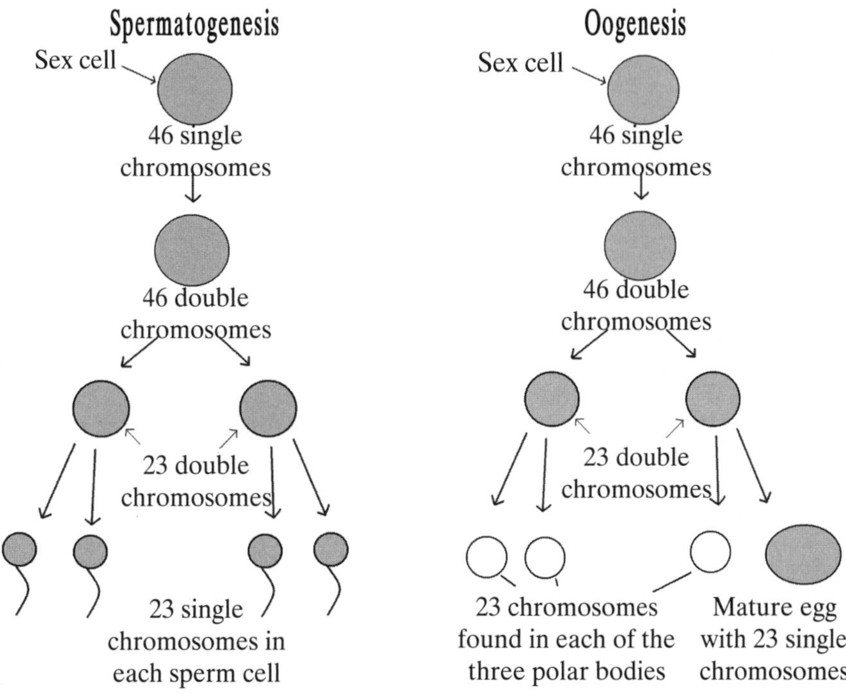

As you can see, one mature male sex cell will produce four viable sperm.

The process in human females where the sex cells divide and mature into haploid egg cells is called oogenesis. A human female will normally produce one mature egg cell about every month from puberty to menopause.

You can see that the development of one female sex cell will produce only one mature egg cell.

Uncontrolled Cell Growth: Cancer

The life span of cells in the human body depends on the type of cell in question. Some cells, such as the ones found on the lining of the esophagus, live only about two days, while there are some types of white blood cells that can live for 20 years. The regulation of growth and the life span of cells is a critical process within the human body. Cells are programmed to reproduce at certain intervals and live for specific periods of time. Most cells are programmed not to reproduce when they are in contact with other cells. Damage to the body usually stimulates cells at the fringes of the injury to begin dividing to help heal the wound. Unfortunately, the growth regulators of the cell sometimes do not work properly. ***Cancer*** is when cells grow and reproduce uncontrollably. When cells reproduce without control, they grow on top of each other forming masses of cells called ***tumors***. Tumors can block and interfere with organ systems within the body, possibly causing imbalance to homeostasis, and death. The more scientists learn about how cells reproduce and divide, the closer we will be to answers for the cure to cancer.

Limits to Cell Size

Scientists observe and study cells under microscopes. The reason for this is that most cells are very small. Table 3.3 gives the size of some various cell types.

Size of Some Various Cells

Cell Type	Size
Red blood cell	20 micrometers
Staphylococcus bacterial cell	2 micrometers
White blood cell	20 micrometers
Plant cell	100 micrometers
Human egg cell	100 micrometers

Table 3.3

There is a very important reason why you do not see any very large cells. Cells have a limit to how large they can grow. The limit to their size is determined by their ability to move materials at proper rates to all the organelles within their cytoplasm. Remember that primarily the process of diffusion conducts the transport of materials within a cell, and diffusion is a fairly slow process. If a cell's volume is close to the surface area of its cell membrane, the diffusion of materials from the environment into the cell and throughout the cytoplasm of the cell can be accomplished. But as the size of a cell increases, its volume increases much quicker than the surface area of the cell membrane. If the volume of the cell was to grow too fast compared to the size of its surface area, then the ability to bring materials into the cell and distribute these materials throughout the cell's cytoplasm would be so greatly reduced that the needed materials would not get to their proper locations within the cell in time. Also, the waste materials produced inside of the cell would take much longer to be excreted. Table 3.4 on page 87 shows the mathematics behind the surface area to volume ratio of a cube. These numbers can be used as a close comparison to a three-dimensional cell.

The table shows how quickly the ratio of surface area to volume decreases as the cell gets larger. A smaller surface area to volume ratio means that it will take more time to bring the amount of needed materials into the interior of the cell.

Ratio of Volume to Surface Area in a 3-Dimensional Cube

Size of the cell	1 cm × 1 cm × 1 cm	3 cm × 3 cm × 3 cm	5 cm × 5 cm × 5 cm
Surface area formula (1 × w × no. of sides)	$1 \times 1 \times 6 = 6$ cm^2	$3 \times 3 \times 6 = 54$ cm^2	$5 \times 5 \times 6 = 150$ cm^2
Volume area (1 × w × h)	$1 \times 1 \times 1 = 1$ cm^3	$3 \times 3 \times 3 = 27$ cm^3	$5 \times 5 \times 5 = 125$ cm^3
Ratio of surface area to volume	6/1 or 6:1	54/27 or 2:1	150/125 or 6:5

Table 3.4

Practice Section 3-4

1. During the cell cycle, DNA is replicated during
   a) G1 b) G2 c) mitosis d) S
2. The first phase of mitosis is called
   a) interphase b) anaphase c) telophase d) prophase
3. The process of mitosis ensures an equal distribution of
   a) cell membranes d) water
   b) chromosomes c) food
4. A disorder in the body where some cells lose the ability to control growth.
   a) angina b) cancer c) diabetes d) malaria
5. The haploid number of chromosomes is equal to
   a) the full set of chromosomes in an organisms
   b) half of the full set of chromosomes in an organism
   c) double the full set of chromosomes in an organism
   d) the number of chromosomes in a normal body cell
6. As the size of the cell increases
   a) the volume increases at the same pace as the surface area
   b) the volume increases faster than the surface area
   c) the surface area increases faster than the volume
7. What is mitosis?
8. What are the three periods of time into which the process of interphase is divided?
9. How is the process of meiosis different from mitosis?

Lesson 3-5: Energy in the Cell

Life's Battery: ATP

Energy is needed by all living organisms to conduct the necessary chemical processes of life. The basic structure of life is the cell, which means all cells must obtain a steady source of energy to survive. The energy source most available for use by the many parts of the cell comes from the molecule called **ATP**. ATP (adenosine triphosphate) is an adenosine molecule chemically bonded to three phosphate molecules.

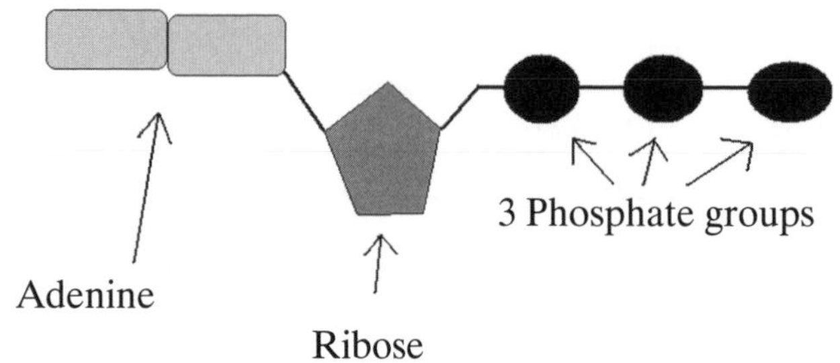

The chemical bonds connecting each phosphate molecule to the adenosine molecule hold energy. Energy is stored when bonds are formed to connect a phosphate molecule to an adenosine molecule, and when these bonds are broken the energy is released. The amount of energy released from these bonds is just the right quantity for most of the chemical reactions that take place within the cell. It is similar to the different size batteries supplied for the many electronic devices on the market today. For example, if you have a CD player that needs AA batteries as an energy source, having D batteries will not solve your problem. Even though the D batteries are a source of energy, this energy is not in the right form or amount to run the device in question. ATP is like the perfect "battery" for most cell activities. When a cell needs an ATP molecule's energy, the last phosphate group breaks off and the energy in the bond is released for use by the cell. ATP molecules bind to the chemicals within the cell that need the energy. This ensures that the energy is transferred from the ATP molecule to the proper place for use within the structure. It is like saying you need to place the battery inside of the CD player to allow the CD player to access the energy of the battery. Once the ATP molecule has released energy by breaking off a phosphate molecule, the ATP molecule now has only two phosphate molecules and is called **ADP** (adenosine diphosphate).

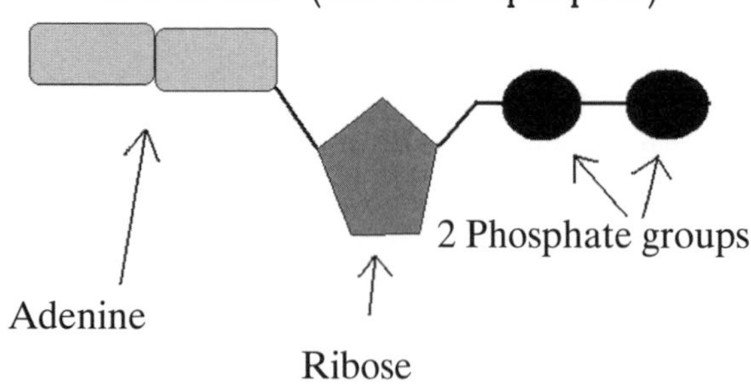

ADP is similar to an uncharged battery because in this state it usually will not provide energy to the cell. The second phosphate molecule connected to adenosine can be removed for energy, but in many cases it does not.

ADP can once again have a third phosphate molecule joined to it and become an ATP molecule. This is why this process is very similar to charging a battery. ADP is like the uncharged battery, and ATP is like a battery charged that is ready to be used for cell activities.

Cellular Respiration

Where does the energy come from that powers the recharging of these very useful ATP molecules? The answer to that question can be found from studying the chemical processes of ***cellular respiration***. Cellular respiration is a chemical process that occurs in all living cells when the trapped energy in the bonds of food molecules is converted to the stored energy in ATP molecules.

The chemical process of cellular respiration starts when a food molecule (glucose) enters the cell and is acted upon by enzymes in the cytoplasm of the cell. A small amount of energy is needed to get the reaction going, and two ATP molecules supply this energy. In this initial reaction a glucose molecule is broken up into two new molecules called pyruvic acid. Chemically, it is said that the glucose molecule has been **oxidized**, because it has lost hydrogen atoms during the process. These four hydrogen atoms are picked up by specific coenzymes called **NAD**, which will take the hydrogen atoms and some high-energy electrons associated with them to other chemical reactions later in the process. NAD becomes NADH when it is holding a hydrogen atom. NAD is said to be a carrier of energy within this system.

Glycolysis

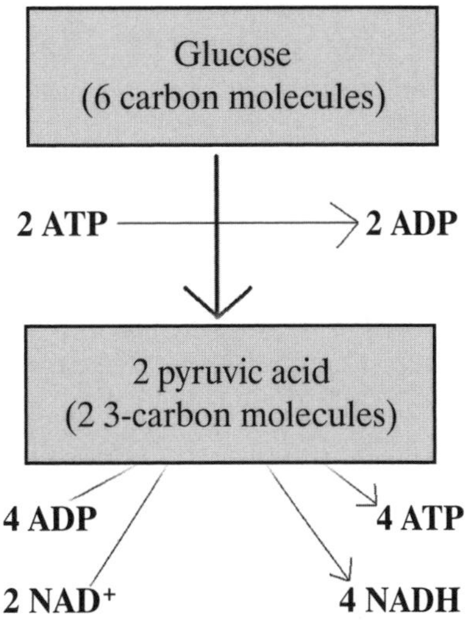

Glucose
(6 carbon molecules)

2 ATP → **2 ADP**

2 pyruvic acid
(2 3-carbon molecules)

4 ADP → **4 ATP**

2 NAD$^+$ → **4 NADH**

Energy is then released as the bonds of the glucose molecule are broken to produce the two pyruvic acid molecules. This energy is stored in the form of four ATP molecules. The net gain of ATP molecules at this point is two, because even though four ATP molecules are produced, two were used to help start the reaction. These initial chemical reactions of cellular respiration are called **glycolysis**. The chemical reactions of glycolysis are considered to be an **anaerobic process**, because they occur in the absence of oxygen. After the process of glycolysis has occurred, only about 10% of all the available energy within the glucose molecules has been used. For some organisms, such as bacteria, this is enough. But for many of the more complex, energy-hungry organisms, such as humans, more energy must still be extracted from the food molecules. Think of the glucose molecule as being a log. Glycolysis is like burning only 10% of the log for heat energy. The aerobic process of respiration that will now be described will look to burn more of the log for energy.

Aerobic Respiration

Which chemical pathway the process of cellular respiration takes after glycolysis will depend on the availability of oxygen and the particular organism (enzymes) in question. Aerobic respiration can occur when oxygen is available for the correct organism. The process of aerobic respiration occurs in two steps: the Krebs cycle and the Electron transport chain.

Krebs cycle

The products of glycolysis are the two pyruvic acid molecules, the NADH coenzyme, and the net gain of two ATP. These products all enter the organelle within the cell called the mitochondria, where the necessary enzymes are found to complete the respiration process. The Krebs cycle is named after the scientist Sir Hans Krebs of England who first discovered many of the chemical processes of this part of the respiration reaction. The Krebs cycle can be summarized in the following manner:

Step 1: The pyruvic acid molecules each have one of their carbon atoms removed in the form of CO_2. The other two atoms of carbon, from the pyruvic acid molecule, are joined to a compound called ***coenzyme A*** to form acetyl-CoA. The two acetyl-CoA molecules are then added to a 4-carbon molecule to make the 6-carbon compound called ***citric acid***.

Step 2: The citric acid produced is broken down into a 4-carbon atom that will eventually be used for the next acetyl-CoA molecules that enter the reaction.

Step 3: The coenzymes NAD and FAD (another similar coenzyme) remove more hydrogen atoms and high-energy electrons from the broken citric acid molecule. These high-energy electrons and hydrogen atoms will be brought to the next phase of this process called the electron transport chain.

Step 4: Two more ATP molecules of stored energy are also produced for use by the cell.

Electron transport chain

The electron transport also occurs inside the mitochondria, actually on the walls of the inner membrane of this organelle where the necessary enzymes are present. The enzymes NADH and FADH found in the inner membrane of the mitochondria pass the energized hydrogen atoms from protein molecule to protein molecule. This process of passing the hydrogen atoms from protein to protein slowly releases the energy inside the hydrogen atoms. Some of this energy is used to create more ATP molecules. Some of this energy released from the hydrogen atoms is also used to create an electrical and chemical gradient from one side of the inner mitochondrial membrane to the outer. This gradient also produces ATP molecules. When the hydrogen atoms have been drained of energy, they must then be removed from the inner membrane. This is where oxygen finally plays a role in the chemical reactions of cellular aerobic respiration. Oxygen acts as the final acceptor of the hydrogen atoms and forms the molecule water. The water that is removed from the inner membrane is actually "cleaning" the inner membrane of hydrogen so that new hydrogen atoms, which are full of energy, can enter and continue the process of cellular respiration. The electron transport chain produces another 32 ATP molecules during this process.

In the chemical process of glycolysis, 10% of the stored energy in the glucose molecule was extracted for cell actvities. The chemical processes of the Krebs cycle and electron transport chain will release another 28% of this stored energy. This means that the total amount of usable energy released from each molecule of glucose during aerobic respiration is just about 38%. This is a very good number when compared to many other energy-extracting machines. Most of the rest of the energy within the molecule will be released as heat.

Fermentation

Sometimes, once the chemical process of glycolysis has ended, there is no available oxygen to begin the aerobic process of respiration. This means that cellular respiration must finish with a different chemical process called fermentation. **Fermentation** is the chemical release of energy from food without the use of oxygen.

Fermentation begins with the process of glycolysis. But at the end of glycolysis, the hydrogen atoms, joined to the NADH molecules, are connected back to the newly formed pyruvic acid molecules. This occurs because the "energy carriers" (NAD) must be available if any type of respiration is to occur. When NAD is in the form NADH it cannot help to initiate the next round of energy-releasing processes in the cell.

Depending on the respiratory enzymes present when the process of fermentation occurs, two pathways are possible after the hydrogen atoms have been added back to the pyruvic acid molecules.

Pathway #1

This pathway usually occurs in smaller, less complex organisms such as yeast cells.

$$\text{pyruvic acid} + \text{NADH} \rightarrow \text{alcohol} + CO_2 + \text{NAD}^+$$

Pathway #2

This pathway usually occurs in larger, more complex organisms such as humans.

pyruvic acid + NADH → lactic acid + NAD^+

It is very important to notice that the chemical processes of fermentation do not oxidize or release any more energy from food than glycolysis. The only real purpose of these secondary pathways is to free up the NAD molecules for continued respiration. Table 3.5 compares the useable energy released from the process of fermentation to the amount of useable energy released from the process of aerobic respiration.

Energy Comparisons Between Aerobic Respiration and Fermentation

Type of Respiration	Number of ATP Produced	Percent of Useable Energy Extracted From Glucose
Aerobic Respiration	36	38%
Fermentation	2	10%

Table 3.5

Looking at the chart, you can see that the process of aerobic respiration is 18 times more efficient at producing ATP (energy) than the process of fermentation.

When do humans undergo the process of fermentation?

Given the previous information, it may seem unusual to that, at times, the human body undergoes the process of fermentation. If human beings are able to achieve the chemical process of aerobic respiration, why would their bodies ever have the need to do the less efficient process of fermentation? Let us look at a possible example of when and how this would occur in the human body.

Example of fermentation in humans

A person stands at the start line of a 100-meter race. In that person's body there is plenty of oxygen in all the cells of the muscles of the legs and arms. At this time, the person is undergoing the chemical process of aerobic respiration.

Then the person runs the 100-meter race and uses large amounts of energy in the process. So much energy, in fact, will be used during this activity that the aerobic respiration process will not be able to release enough energy from food molecules in the body quickly enough to supply all that is needed. The human body is not able to breathe in enough oxygen to extract all the hydrogen atoms that have quickly accumulated in the mitochondria organelles in the muscle cells. Many of the aerobic chemical processes of respiration begin to shut down in these muscle cells because of the excess low-energy hydrogen atoms located there.

Fermentation now begins in the human body. The only way these muscle cells can now obtain energy is by releasing energy from food without using oxygen in the process of fermentation. The problem with the process of fermentation is that it produces much less energy than aerobic respiration. The chemical process of fermentation also produces as a by-product the chemical lactic acid that builds up in the muscle cells causing them to burn and cramp up. This means that at the end of the race, because of the switch in respiration processes that occurred in the muscle cells of the arms and legs, the person will need to breathe very heavily to obtain enough oxygen to clean the low energy hydrogen atoms from the mitochondria and to also clean the lactic acid out of the muscle cells of the legs and arms. Usually in about 5 to 10 minutes after such an activity the heavy breathing has accomplished both of these tasks.

••••••

Practice Section 3-5

1. During the process of respiration, energy from the oxidation of food is stored in
   a) DNA b) PGAL c) RNA d) ATP
2. What organelle inside the cell is most closely associated with the chemical process of respiration?
   a) lysosome b) mitochondria c) ribosome d) nucleus
3. The production of lactic acid occurs during which chemical process?
   a) aerobic respiration c) photosynthesis
   b) anaerobic respiration d) Kreb's cycle
4. Which of the following chemical processes releases the most energy in the form of ATP?
   a) glycolysis c) electron transport chain
   b) Krebs cycle d) fermentation
5. Where in the cell does the chemical process of glycolysis occur?
   a) mitochondria b) cell wall c) cytoplasm d) ribosome
6. Which of the following molecules is considered to be an electron acceptor?
   a) PGAL b) RNA c) NAD d) ATP
7. What are the two types of cellular respiration?
8. Fermentation is a chemical process that is carried out in an organism under what conditions?
9. What is ATP? What is its significance in biology?
10. What is the difference in production of ATP molecules between the processes of aerobic respiration and anaerobic respiration?

Lesson 3-6: Autotrophic Cells: Photosynthesis

There are many cells on Earth that are able to manufacture organic molecules from inorganic molecules. These organisms that are able to make their own food are said to be **autotrophic**. All of these organisms that manufacture their own food need an energy source in order to accomplish this task. Most organisms on Earth that make their own food will use the sunlight as the energy source, and we will discuss these types of food-making processes in just a second. There is also a group of organisms that can obtain energy from the chemical bonds of inorganic molecules to manufacture food in a process called ***chemosynthesis***. Many of these chemosynthetic organisms dwell in places where there is no sunlight to use as an energy source, such as deep under the ocean waters and under the mud in swamps.

Many other organisms use the most abundant energy source on this planet, the sun, to create food. Cells from organisms that use the sun as an energy source—plants, algae, some bacteria, and some protests—contain the green pigment chlorophyll and are able to create organic molecules (food) from inorganic molecules in the chemical process known as **photosynthesis**.

Example

The following reaction summarizes the chemical process of photosynthesis:

carbon dioxide + water → glucose + oxygen + water

$$6CO_2 + 6H_2O \rightarrow C_6H_{12}O_6 + 6O_2 + 6H_2O$$

The first equation shows the names of the chemicals involved in the reaction. Carbon dioxide and water are said to be the reactants in the equation. Glucose, oxygen, and water are said to be the products in the equation. The second equation is the molecular formula of each chemical and the number of molecules needed of each chemical for the reaction to occur.

The chemical processes of photosynthesis can be broken into three stages:

1. The capture of light energy.
2. The conversion of light energy into chemical energy.
3. The storage of chemical energy in sugar.

••••••

The first two steps listed are sometimes called the ***light reaction*** of photosynthesis, while the third step is called the ***dark reaction*** of photosynthesis.

1. The capture of light energy

White light is a form of radiant energy made up of seven different wavelengths of light called the visible spectrum (red, orange, yellow, green, blue, indigo, and

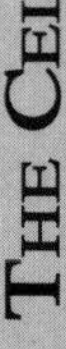

violet energy). Red light has the longest wavelength and the least amount of energy, and violet light has the shortest wavelength and the largest amount of energy.

Chlorophyll, the green pigment found in food-producing cells, is a very complex molecule made of more than 120 atoms. The arrangement of all these atoms in the molecule help chlorophyll trap certain wavelengths of light energy. Red light energy and blue light energy are trapped the most efficiently, while green light energy is trapped the least. This is why chlorophyll appears green, because this light energy cannot be trapped and is reflected back into your eye from the molecule. Chlorophyll is found inside organelles called ***chloroplasts***.

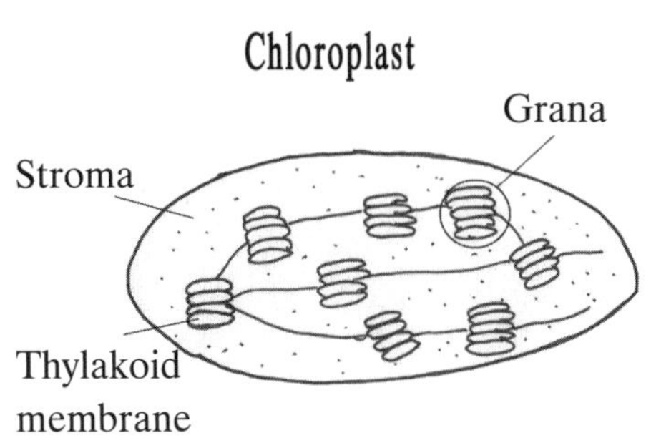

Chloroplasts are double-membrane bound, oval-shaped organelles found in the cells of autotrophic organisms. A leaf cell has an average of about 50 chloroplasts in its cytoplasm. Each chloroplast contains all the chlorophyll and enzymes needed to complete the complex chemical reactions of photosynthesis. There are inner double-membrane bound structures called ***thylakoids*** that are stacked in structures called ***grana*** located throughout the interior of the chloroplast. The rest of the interior of the chloroplast is filled with a protein liquid called ***stroma***. The chlorophyll traps the light energy in the electrons of its atoms. Electrons that receive energy from outside the atom move to a higher electron shell around the nucleus, creating a form of potential energy. This stored potential energy is realized when the electrons fall back to their original shells and release the energy.

2. The conversion of light energy into chemical energy

The energy released from the electrons in the chlorophyll is used to do two things: (1) split water molecules and (2) make ATP.

When the water molecule is split in the chemical process called ***photolysis***, hydrogen ions (H^+) and hydroxide ions (OH^-) are formed. Oxygen is released from the hydroxide ion into the environment. The hydrogen ions as well as high-energy electrons associated with them are collected by a coenzyme called NADP. NADP brings these hydrogen ions and electrons, which have potential energy in them, to the next step of the photosynthesis process.

Some of the energy from the electrons in chlorophyll is used to change ADP into ATP. This ATP will also be used in the next step of photosynthesis.

So you can see in this second step that the light energy is converted to chemical energy in the form of ATP and the ions of hydrogen are held in place in the coenzyme NADP.

3. **Storing chemical energy in sugar molecules**

This third stage of photosynthesis is sometimes referred to as the "dark reaction." It is not called the dark reaction because it occurs at night, but because the steps of this process do not involve the absorption of light energy. The dark reaction is also sometimes called the Calvin cycle, named after Melvin Calvin who discovered many of the steps of this process. In the dark reaction, carbon dioxide, which is brought into the photosynthetic organism from the environment, is chemically reduced by the hydrogen ions produced in the earlier steps of this process. The following is an unbalanced molecular equation of this last step.

$$CO_2 + H \rightarrow CH_2O$$

In the equation, the molecule CH_2O represents the carbohydrate produced from this reaction. The name of the first real carbohydrate produced is a 3-carbon molecule called PGAL (phosphoglyceraldehyde). PGAL is then used to make all the other macromolecules that are produced.

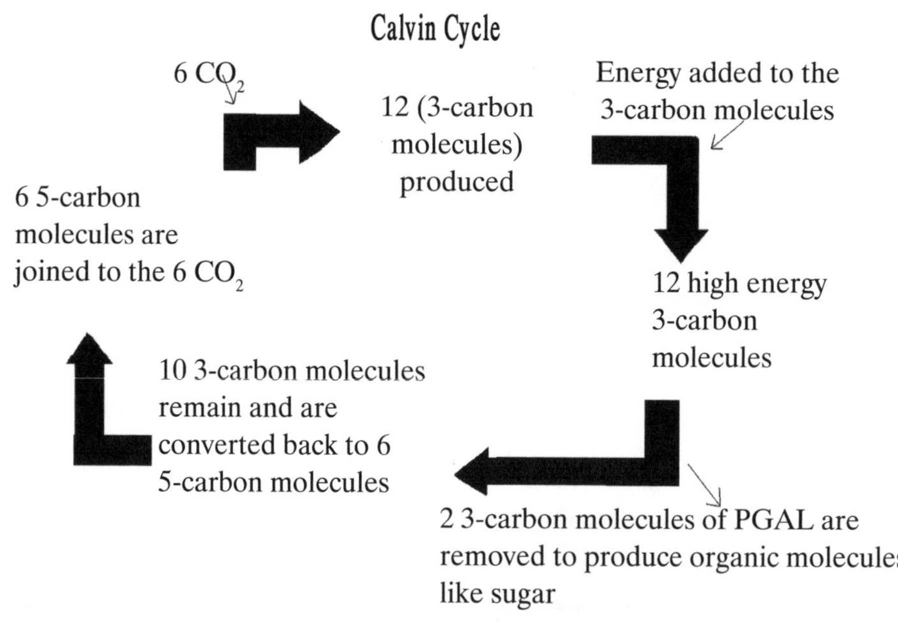
Calvin Cycle
6 CO2
12 (3-carbon molecules) produced
Energy added to the 3-carbon molecules
12 high energy 3-carbon molecules
2 3-carbon molecules of PGAL are removed to produce organic molecules like sugar
10 3-carbon molecules remain and are converted back to 6 5-carbon molecules
6 5-carbon molecules are joined to the 6 CO2

Factors That Affect the Rate of Food Production

The rate of photosynthesis is affected by many factors within the environment, including:

- The wavelengths of light available.
- The temperature.
- The carbon dioxide concentration.
- The amount of chlorophyll available within the organism.

How do you think increasing the temperature would affect photosynthesis? How do you think increasing the carbon dioxide concentrations would affect photosynthesis? How about increasing or decreasing certain wavelengths of light? Think about the steps you have just read concerning the chemical processes of this reaction.

Practice Section 3-6

1. An organism that is not able to manufacture organic molecules from inorganic molecules is
   a) chemosynthetic organism c) autotrophic organism
   b) heterotrophic organism d) photosynthetic organism
2. Which of the following are products of photosynthesis?
   a) carbon dioxide and water c) glucose and oxygen
   b) carbon dioxide and glucose d) oxygen and calcium
3. What is the name of the pigment used to trap light energy during the process of photosynthesis?
   a) PGAL b) ATP c) chlorophyll d) stroma
4. A water molecule is split into hydrogen and hydroxide ions during the process of
   a) chemosynthesis c) photolysis
   b) hydrolysis d) carbon fixation
5. What energy conversion occurs during the process of photosynthesis?
   a) light energy to nuclear energy
   b) chemical bond energy to light energy
   c) light energy to chemical bond energy
   d) mechanical energy to radiant energy
6. The coenzyme that carries hydrogen atoms from the light reactions to the dark reactions is
   a) ATP b) chlorophyll c) NADP d) RNA
7. What is the chemical equation of photosynthesis?
8. Which is the cell organelle responsible for the absorption of light during the photosynthetic process?
9. How does the chlorophyll "trap" light energy?
10. What is the purpose of photosynthesis?

Terms From Chapter 3			
active transport	chromosome	Golgi bodies	NADP
anaphase	cilia	interphase	organelle
cell membrane	crossing over	lysosome	PGAL
cell nucleus	cytoplasm	meiosis	passive transport
cell theory	diffusion	metaphase	prophase
cell wall	diploid	microtubules	ribosome
centriole	endoplasmic reticulum	mitochondria	telophase
centrosome	flagella	mitosis	vacuole
chlorophyll	gamete	monoploid	

Chapter 3: Exam

Matching Column for Cell Organelles

Match the cell organelle with the characteristic that best describes that organelle. You may use a choice more than once or not at all.

Cell Organelle

1. ribosome
2. mitochondria
3. vacuole
4. Golgi body
5. plasma membrane
6. nucleus
7. lysosome
8. cell wall
9. chloroplast
10. endoplasmic reticulum
11. cytoplasm

Characteristics

a. Contains chlorophyll and is only found in plant cells.
b. Controls the transport of materials into and out of the cell.
c. Storage sac for water and wastes in the cell.
d. A sac full of chemicals that are used in cellular digestion.
e. A watery medium in which many of the cell's chemical reactions take place.
f. The information center of the cell that contains DNA.
g. The site of cellular respiration.
h. The site of protein synthesis in the cell.
i. Packages secretions, within the cell, before they leave the cell.
j. Intracellular membranes that help transport materials throughout the interior of the cell.
k. Protects and supports the shape of the plant cell.

Matching Column for Scientists

Match the scientist with their contribution to the field of biology.

Scientist		Contribution to Science
12. Robert Hooke	a.	Built the first simple microscope.
13. Robert Brown	b.	From experimentation concluded all animals are made of cells.
14. Anton van Leewenhoek	c.	From experimentation concluded all plants are made of cells.
15. Rudolph Virchow	d.	First to describe the nucleus of the cell.
16. Matthias Schleiden	e.	Scientist who contributed the concept that "all cells come from other cells."
17. Theodor Schwann	f.	Invented the compound microscope.
18. Zacharias Janssen	g.	First to see bacteria and protozoa under the microscope.

Multiple Choice

19. The source of oxygen produced by photosynthesis is
    a) water b) carbon dioxide c) sugar d) starch
20. Which occurs during the light reaction of photosynthesis?
    a) Water molecules are split.
    b) Carbon dioxide reacts with hydrogen.
    c) PGAL molecules are converted to glucose.
    d) Oxygen is combined with carbon dioxide.
21. Which occurs during the dark reaction of photosynthesis?
    a) Carbon dioxide is united with hydrogen.
    b) Photolysis.
    c) Water is split into hydrogen and oxygen.
    d) ATP is produced.
22. The dark reaction of photosynthesis cannot occur in the absence of
    a) light b) chlorophyll c) CO_2 d) O_2

Short Response

23. What are the major differences between a plant cell and an animal cell?
24. Explain how the chemical processes of photosynthesis and respiration can be considered opposites of each other.
25. What is the significance of sex cells being created by meiosis instead of mitosis?
26. What evidence, based on your knowledge of the chemical processes of respiration, suggests that glycolysis is a more ancient process than aerobic respiration?

Answer Key

Answers Explained Section 3-1

1. **B** is the correct answer because Robert Hooke coined the word “cell” when he observed a piece of cork under the microscope.

   A is not correct because van Leeuwenhoek was the first to see bacterial cells and protozoa under the microscope.

   C is not correct because Brown was the first to identify the nucleus of the cell.

   D is not correct because Schwann discovered that all animals are made of cells.

2. **C** is the correct answer because “all cells contain a nucleus” is not part of the cell theory.

   A, **B**, and **C** are not correct because all these statements are part of the cell theory.

3. **C** is the correct answer because the invention of the microscope led most directly to the formation of the cell theory because all cells must be observed by humans using this instrument.

   A is not correct because an ultracentrifuge spins substances at very high speeds to separate materials based on density.

   B is not correct because a balance is used to measure the mass of an object.

   D is not correct because chromatography is used to separate chemicals based on the rate they are deposited on special paper.

4. **D** is the correct answer because the organelle that contains its own DNA is the mitochondria.

   A, **B**, and **C** are not correct because the cell membrane, ribosomes, and vacuoles are all organelles that do not contain their own DNA.

5. **C** is the correct answer because viruses are not composed of cells. Viruses are DNA coated in protein. This fact leads many scientists to not classify viruses as living organisms.

   A, **B**, and **D** are not correct because plants, bacteria, and animals are all made of cells.

6. **B** is the correct answer because Darwin had many important discoveries in evolution and plant biology, but his work did not help formulate the cell theory.

   A, **C**, and **D** are not correct because Schwann, Virchow, and Schleiden all conducted experiments that led to the formation of the cell theory.

7. The cell theory is the idea that the cell is the basic unit of both structure and function in all living things. This theory also states that all cells most come from pre-existing cells. This theory is the most basic foundation of all of biology.
8. Viruses are not constructed of cells, but are capable of carrying on the life process of reproduction (with the help of cells).
9. Cork is made from tree bark, which means it is constructed of plant cells. Robert Hooke discovered the "plant" cell by looking at these samples and then coined the word "cell" because of what he believed were similarities between plant cells and the shape of prison cells.

Answers Explained Section 3-2

1. **D** is the correct answer because prokaryotic cells do all of the following: grow and reproduce, contain cell membranes and cytoplasm, and lack a nucleus.
2. **C** is the correct answer because animal cells do not contain cell walls.

   A, **B**, and **D** are not correct because bacterial cells, plant cells, and fungi cells all contain cell walls.
3. **D** is the correct answer because the vacuole is the organelle that stores materials for the cells.

   A is not correct because the cell membrane is the semipermeable boundary of the cell.

   B is not correct because the endoplasmic reticulum consists of twisted membranes scattered throughout the inside of the cell which become work sites for proteins synthesis and compartmentalizing chemical reactions.

   C is not correct because the mitochondria are where respiration occurs in the cell.
4. **C** is the correct answer because in the chloroplast, the chemical reaction that occurs is photosynthesis.

   A is not correct because respiration occurs in the mitochondria

   B is not correct because excretion is accomplished in the cell by the lysosomes and the cell membrane.

   D is not correct because chemosynthesis is a chemical process that occurs in bacterial cells that are able to use inorganic chemicals to supply energy for food production.
5. **B** is the correct answer because the cell membrane regulates the movement of materials into and out of the cell. The cell membrane is said to be semipermeable to materials.

A is not correct because the cell wall found surrounding some cells provides structure and support for the cell but cannot regulate materials that enter the cell.

C is not correct because the nucleus contains the DNA of the cell.

D is not correct because the ribosome is the site of protein synthesis in the cell.

6. **D** is the correct answer because the ribosomes are the site of protein synthesis.

   A is not correct because ribosomes do not contain DNA.

   B is not correct because ribosomes are the smallest organelles in the cell.

   C is not correct because mitochondria, not ribosomes, are the site of protein synthesis.

7. The cell membrane forms the outside boundaries of the cell to its environment and because it is selectively permeable regulates the flow of materials into and out of the cell.

8. Ribosomes can be found floating freely in the cytoplasm fluid of the cell and also can be found attached to the endoplasic reticulum found within the cell.

9. Lysosomes perform the action of intracellular digestion through the use of digestive enzymes created by the endoplasmic reticulum.

Answers Explained Section 3-3

1. **D** is the correct answer because **A**, **B**, and **C** are true about the cell membrane: it regulates movement into the cell, it is composed of lipids and proteins, and it surrounds all living cells.

2. **A** is the correct answer because diffusion is the movement of molecules from an area of high concentration to an area of low concentration.

   B is not correct because the movement of molecules from an area of low concentration to an area of high concentration is active transport.

   C is not correct because molecules are always in motion.

   D is not correct because the movement of food into the cell with the expenditure of energy is also active transport.

3. **B** is the correct answer because a solution that is considered hypertonic has more dissolved materials than a hypotonic solution.

   A is not correct because hypertonic solutions always have more, not less, dissolved materials than hypotonic solutions.

   C is not correct because two solutions with the same amount of dissolved materials are called isotonic solutions.

4. **C** is the correct answer because active transport requires the expenditure of cellular energy.

   A is not correct because osmosis is the diffusion of water.

   B is not correct because diffusion is the movement of materials without the expenditure of cellular energy.

   D is not correct because diffusion is a type of passive transport.

5. **B** is the correct answer because the cell membrane's bilayer is made up of phospholipids.

   A is not correct because protein molecules are part of the cell membrane but embedded in the phospholipid bilayer.

   C is not correct because carbohydrates are present on the surface of some cells, but do not constitute the structure of the lipid bilayer.

   D is not correct because nucleic acids are found primarily in the nucleus of the cell.

6. **A** is the correct answer because the action of the infolding of the cell membrane to allow the intake of large molecules is called endocytosis.

   B is not correct because exocytosis is the removal of large materials from the cell from the fusion of the vacuole with the cell membrane.

   C is not correct because osmosis is the diffusion of water molecules through the cell membrane.

   D is not correct because diffusion is the passive transport of materials through the cell membrane.

7. The two main components of the cell membrane include phospholipids and proteins. The phospholipids form the lipid bilayer which forms the surrounding structure of the cell and the proteins are embedded in the lipid layer to aid in transport of materials into and out of the cell.

8. Passive transport is the movement of molecules from an area of high concentration to an area of low concentration. No additional energy must be added to the system for this to occur, the natural movement of molecules is all that is needed. Active transport is the movement of molecules from an area of low concentration to an area of high concentration. This movement is against normal molecular motion, so energy must be added to the system for this to occur.

9. Endocytosis is a process where the cell membrane surrounds a sample of food engulfing it within the body of the cell. Because this process of engulfing the food item requires energy it is considered a form of active transport.

Answers Explained Section 3-4

1. **D** is the correct answer because the DNA is replicated during the S phase of the cell cycle.

   A is not correct because G1 is the phase of the cell cycle where cell growth occurs.

   B is not correct because G2 is the phase of the cell cycle where the cell prepares chemically for cell division.

   C is not correct because mitosis is stage of the cell cycle where the chromosomes and cell separate.

2. **D** is the correct answer because the first phase of mitosis is called prophase.

   A is not correct because interphase is not a phase of mitosis, it is where G1, S, and G2 occur.

   B is not correct because anaphase is the third phase of mitosis where the chromosomes actually separate.

   C is not correct because telophase is the last phase of mitosis where the two newly formed cells begin to separate.

3. **B** is the correct answer because the process of mitosis ensures the equal distribution of chromosomes to each of the newly formed daughter cells.

   A is not correct because the cell membranes of the newly created daughter cells are actually divided sections of the parent cell membrane. Many times the cell membrane is not divided equally.

   C and **D** are not correct because food and water are separated when the cytoplasm is divided during cytokinesis after mitosis is over. The distribution of food and water is not even.

4. **B** is the correct answer because cancer is a condition where cells in certain parts of an organism lose the ability to control cell growth.

   A is not correct because angina is a condition where a region of the heart muscle receives an insufficient supply of blood.

   C is not correct because diabetes is a condition where the pancreas does not produce the correct levels of insulin.

   D is not correct because malaria is a disease caused by the protist *Plasmodium* and is spread by the *Anopheles* mosquito.

5. **B** is the correct answer because the haploid number of chromosomes is equal to half of the full set of chromosomes in an organism.

   A is not correct because the full set of chromosomes is called the diploid number.

C is not correct because double the full set of chromosomes in an organism would be called the tetraploid number.

D is not correct because the number of chromosomes in a normal body cell is said to be the diploid number.

6. **B** is the correct answer because as the size of the cell increases, the volume of the cell increases faster than the surface area increases. This condition makes it harder for the cell to diffuse materials taken in from the environment and thus limits possible cell size.

   A is not correct because the volume does not increase at the same pace as the surface area.

   C is not correct because the surface area does not increase faster than the volume.

7. Mitosis is the chemical process where a eukaryotic cell divides into two new similar cells. The process of mitosis is the basic process responsible for embryonic development and tissue regeneration in multi-cellular organisms. Mitosis is also the manner in which many one celled organisms reproduce asexually.

8. Interphase is the period of time that occurs in the cell before cell division (mitosis). Interphase consists of three periods called G1, S, and G2. The G1 phase is the normal growth period of the cell. The S phase is when the genetic information inside the cell is replicated. The G2 phase is when the cell is chemically preparing itself for the process of mitosis.

9. Mitosis is the process of cellular replication for the growth and development of multicellular organisms and for asexual reproduction in one-celled organisms. Meiosis is the process where the genetic information is reduced in half inside the (sex) cell in preparation for the possible union with another (sex) cell during the process of fertilization. Meiosis is critical for the process of sexual reproduction.

Answers Explained Section 3-5

1. **D** is the correct answer because the energy released from the process of respiration is stored in a molecule of ATP.

   A is not correct because DNA is the hereditary chemical of life.

   B is not correct because PGAL is the first organic compound produced from photosynthesis.

   C is not correct because RNA is a nucleic acid that aids in protein synthesis.

2. **B** is the correct answer because the mitochondria are the site of cellular respiration.

A is not correct because the lysosome is an organelle that carries out cellular digestion.

C is not correct because the ribosome is the site of protein synthesis in the cell.

D is not correct because the nucleus contains the DNA and RNA of the cell.

3. **B** is the correct answer because lactic acid is produced during anaerobic respiration.

   A is not correct because aerobic respiration produces carbon dioxide and water as products.

   C is not correct because photosynthesis produces sugar, oxygen, and water as products.

   D is not correct because the Krebs cycle is a part of the process of aerobic respiration.

4. **C** is the correct answer because the electron transport chain releases the most energy in the form of ATP. The number of ATP produced is 32.

   A is not correct because glycolysis is an aerobic first phase of cellular respiration that produces a net gain of two ATP.

   B is not correct because the Krebs cycle is a part of the aerobic respiration process and produces two ATP.

   D is not correct because fermentation is a form of anaerobic respiration and produces a net gain of two ATP.

5. **C** is the correct answer because the chemical process of glycolysis occurs in the cytoplasm of the cell.

   A is not correct because the mitochondria is where the Kreb's cycle and the electron transport chain of aerobic respiration occurs in the cell.

   B is not correct because the cell wall is an organelle that helps support the cells of plants, fungi, and bacteria.

   D is not correct because the ribosome is an organelle in the cell that is the site of protein synthesis.

6. **C** is the correct answer because NAD is considered to be an electron acceptor in the chemical process of glycolysis.

   A is not correct because PGAL is the first organic compound produced in the process of photosynthesis.

   B is not correct because RNA is a nucleic acid that aids in protein synthesis in the cell.

   D is not correct because ATP is a chemical found in living organisms used to store and release energy.

7. The two types of cellular respiration are aerobic cellular respiration (a chemical reaction that uses molecules of oxygen (O_2) to produce energy, and anaerobic cellular respiration (a chemical reaction that does not use molecules of oxygen (O_2) to produce energy.
8. Fermentation is a chemical process that obtains energy for the organism when there are no oxygen molecules available. Fermentation is considered to be a type of anaerobic respiration.
9. ATP is adenosine triphosphate an important carrier of energy in all living cells. ATP is formed from adenosine diphosphate (ADP) by an oxidation reaction in mitochondria, or by a photo reaction in plants. ATP is important because it is the universal energy currency for metabolism. ATP is the perfect amount of energy for cells to operate.
10. In anaerobic respiration two ATP molecules are created from the breakdown of one glucose molecule. In aerobic respiration 36 ATP molecules are created from the breakdown of one glucose molecule. This obviously means that aerobic respiration is a much more efficient energy obtaining biological mechanism.

Answers Explained Section 3-6

1. **B** is the correct answer because an organism not able to manufacture organic molecules is called a heterotroph.

   A is not correct because an organism that is chemosynthetic is able to manufacture its own organic molecules by using the energy found in organic molecules.

   C is not correct because an autotrophic organism can manufacture its own organic molecules by either photosynthesis or chemosynthesis.

   D is not correct because a photosynthetic organism can manufacture its own organic molecules by trapping and using light energy.
2. **C** is the correct answer because the products of photosynthesis are oxygen and glucose.

   A is not correct because carbon dioxide and water are the products of aerobic respiration.

   B and **D** are not correct because they not products of respiration or photosynthesis.
3. **C** is the correct answer because the chemical used to trap light energy during the process of photosynthesis is chlorophyll.

   A is not correct because PGAL is the first organic compound produced during the process of photosynthesis.

B is not correct because ATP is a chemical found in living organisms used to store and release energy.

D is not correct because the stroma is the region inside the chloroplast of the cell.

4. **C** is the correct answer because photolysis is the splitting of a water molecule.

   A is not correct because chemosynthesis is the manufacture of organic molecules by using the chemical energy located in inorganic molecules.

   B is not correct because hydrolysis is the splitting of two molecules with the addition of a water molecule.

   D is not correct because carbon fixation is the chemical process where atmospheric carbon dioxide is captured by plants.

5. **C** is the correct answer because the energy conversion that occurs during the chemical process of photosynthesis is light energy to chemical bond energy.

   A, **B**, and **D** are not correct because they are energy conversions that do not occur in the process of photosynthesis.

6. **C** is the correct answer is NADP is the coenzyme that carries hydrogen atoms from the light reaction to the dark reaction.

   A is not correct because ATP is a chemical found in living organisms used to store and release energy.

   B is not correct because chlorophyll is the chemical in photosynthetic plants used to trap light energy for food production.

   D is not correct because RNA is a nucleic acid that aids in protein synthesis.

7. $6\,CO_2 + 6\,H_2O \rightarrow C_6H_{12}O_6 + 6\,O_2$ Trapped light energy is the powerhouse that drives this chemical reaction.

8. Light energy is absorbed (trapped) by the chemical liquid chlorophyll which is found inside the chloroplast organelle.

9. Light energy excites electrons inside the chlorophyll molecules. These excited electrons liberate energy which is used to form ATP molecules from ADP molecules. It is within these ATP molecules that the energy is stored for the production of food in photosynthesis.

10. Photosynthesis is a chemical process that takes place in plants, algae, some bacteria and some protists, that produce usable sugar "food energy" for the cells of the organisms.

Answers Explained Chapter 3 Exam

Matching Column

1. H	4. I	7. D	10. J	13. D	16. C
2. G	5. B	8. K	11. E	14. G	17. B
3. C	6. F	9. A	12. F	15. E	18. A

Multiple Choice

19. **A** is the correct answer because in the chemical reaction of photosynthesis molecules of water are split in the process of photolysis. The resulting oxygen molecules from this reaction are deposited into the environment.

    B is not correct because even though carbon dioxide is taken in and used during the photosynthesis reaction, oxygen is never released from these molecules.

    C and **D** are not correct because these are products resulting from the reaction of photosynthesis, which means the oxygen molecules present in these compounds are chemically bonded in place.

20. **A** is the correct answer because water is split during the process of photolysis, during the light reaction of photosynthesis, to produce more needed hydrogen atoms and electrons.

    B is not correct because hydrogen reacts with carbon dioxide in the dark reaction of photosynthesis.

    C is not correct because PGAL is converted to glucose in the dark reaction of photosynthesis.

    D is not correct because this reaction does not occur anywhere during the chemical process of photosynthesis.

21. **A** is correct because carbon dioxide reacts with hydrogen to form organic compounds during the dark reaction of photosynthesis.

    B is not correct because the chemical process of photolysis (water splitting) occurs during the light reaction of photosynthesis.

    C is not correct for the same reason **B** is not correct.

    D is not correct because all the ATP produced occurs in the light reaction of photosynthesis.

22. **C** is the correct answer because carbon dioxide is needed to complete the synthesis of organic molecules during the dark reaction of photosynthesis.

    A is not correct because light is needed to complete the light reaction, not the dark reaction of photosynthesis. It is important to note here that the dark reaction would eventually cease if the light reaction was unable to occur.

B is not correct because chlorophyll is needed in the light reaction to help trap light energy to power the chemical reaction.

D is not correct because oxygen is actually produced during the light reaction of photosynthesis and not used in the dark reaction.

Short Response

23. The following is a list of the major differences between a plant and an animal cell.
    - An animal cell has no cell wall or chloroplasts.
    - An animal cell does have a centriole and very small vacuoles.
    - A plant cell has chloroplasts and a cell wall.
    - A plant cell does not have a centriole and has very large vauoles.
24. The chemical reactions of photosynthesis and respiration are considered to be opposite processes of each other because the reactants of photosynthesis are the products of respiration, and the reactants of respiration are the products of photosynthesis.

    Respiration:

    $$O_2 + C_6H_{12}O_6 \rightarrow CO_2 + H_2O + \text{energy}$$

    Photosynthesis:

    $$CO_2 + H_2O + \text{energy} \rightarrow O_2 + C_6H_{12}O_6$$
25. The creation of sex cells by the process of meiosis is important because during the process of meiosis, the number of chromosomes in the parent cell is reduced to half in the daughter cells. Sex cells must contain half the number of chromosomes in order for sexual reproduction and the union of two sex cells to be successful.
26. Glycolysis must be a more ancient process than aerobic respiration because glycolysis occurs at the beginning of every type of chemical process of respiration on this planet, including aerobic respiration.

Genetics

The branch of biology that studies the process of heredity is called **genetics**. ***Heredity*** is the ability of a living organism to pass on information about itself to its offspring concerning its physical and chemical make-up. When an organism reproduces **asexually**, a one-parent organism produces genetically identical offspring. The offspring are identical to the parent because all the genes from that parent end up in the offspring. When an organism reproduces **sexually**, genes from both parents help determine the characteristics of the offspring. In both of these types of reproduction, the offspring tend to resemble the parent or parents because of this transfer of genes.

Lesson 4-1: The Work of Gregor Mendel

The scientific study of genetics started with the work of an Austrian monk by the name of Gregor Mendel (1822–1884). Mendel did extensive work concerning the transmission of traits from parent organisms to offspring on several plant species. But it is the genetic work that Gregor Mendel did on the pea plants at his gardens in the monastery that would become the foundation of modern genetics.

Pollination of the Garden Pea Plant

The garden pea plants that Mendel used in his experiments reproduce sexually. ***Pollen***, which is the male gamete of the plant, and the **ovules**, which are the female gametes of the plant, are both found inside the same flower. This means that the plant ***self-pollinates*** (fertilizes itself) to produce the seeds of the next generation. This worked to Mendel's advantage in many of his experiments. Occasionally, Mendel needed to cross two different pea plants; in botany, this is called **cross-pollination.** To achieve cross-pollination of the pea plant, Mendel removed the part of the plant that contains the pollen, called the ***anther***, from the flower. Removal of the anther would not allow the flower to self-pollinate. Mendel could then take pollen from a different plant and mix the new pollen with the ovule of the plant that had its anther removed.

Pea Plant Experiment

The beginning of Mendel's famous pea plant experiments start with the cross-pollination of tall pea plants with small pea plants. The tall plants and the small plants crossed in the initial part of the experiment were "true breeds" for height. This means that all previous self-pollinating experiments conducted with the tall pea plants had always produced only tall pea plants. The same was true for the small pea plants, every time a small pea plant was crossed with a small pea plant, it only produced small pea plants. Both the tall and short plants are considered to be **homozygous** for their characteristic of height. This means that each plant has two of the same genes for height: the small plant with a small gene for height found on each chromosome and the tall plant with a tall gene for height found on each chromosome. The alternate genes for a trait found on each chromosome are called **alleles**. The first generation of Mendel's study crossing "true breeding" or ***pure*** tall pea plants with pure short pea plants is also sometimes called the ***parent***, or P1 generation. The results of the first part of the plant breeding experiment shocked Mendel. The cross between the tall and short pea plants produced seeds that only grew into tall pea plants. It seemed that the trait for shortness in pea plants had disappeared. Mendel then crossed-pollinated all the tall pea plants produced from the parent generation. The tall plants that were crossed were called the ***first filial*** (F1) generation. When Mendel planted the seeds produced from the F1 generation, about 75% of them grew into tall pea plants and about 25% of them grew into short pea plants. It now appeared that the trait for shortness in pea plants had returned in the F2 generation!

These experiments led Mendel to conclude three very important facts about the transmission of traits:

1. **The Principle of Dominance**

   The cross between the pure tall plants with pure short plants led to all tall offspring. In this experiment, each plant gave the offspring its trait for height, a tall trait from the tall plant, and a short trait from the short plant. But the appearance of the offspring only showed the tall trait. Mendel concluded that in a pea plant that has a tall trait and a short trait, the plant would only display the tall trait. How the organisms displays or looks because of its genes is called its **phenotype**. The genetic make-up of the organism is called its **genotype**. The tall trait is said to be the **dominant trait**, because it is the trait that determines the phenotype, while the short trait in the pea plants is said to be the **recessive trait** because it is present in the genotype but does not show up phenotypically unless it is the only gene present.

2. **The Law of Segregation**

   The Law of Segregation means that gametes, or sex cells, only contain one gene for each trait. This means that the offspring will receive one gene from each parent for a particular characteristic. If we look at the F1 generation of Mendel's pea plant experiment, the phenotype of the plants that were crossed was tall, but the genotype of the plants was **heterozygous**, meaning one gene for tallness and one gene for shortness. This means that when the F1 plants were crossed to produce the F2 generation, each plant could give either a tall gene or a short gene to the next generation.

3. **Law of Independent Assortment**

   When Mendel actually did his experiment he studied seven different traits about the pea plant and how all of these traits were transmitted to the next generation. What Mendel observed in his experiments was that each trait was transmitted separately to the offspring. For example, if a parent pea plant gives a gene for tallness to one particular offspring, it does not have to necessarily give a gene for green pea color. The gene for the color of the pea is inherited independently from the gene for height. One particular fact about the assortment of genes has been updated since the experiments of Mendel. Genes that are on different chromosomes are always independently assorted to the offspring, but genes on the same chromosome can be assorted independently or can be assorted in groups.

Many times when studying the crosses between two organisms a more organized approach can be taken with the use of a Punnett square. A **Punnett square** is an organized method of predicting the probability of certain genetic outcomes from a given cross between two organisms with known genotypes.

Example

If a heterozygotus tall pea plant is crossed with a homozygous, or pure, short pea plant, what is the expected ratios of the offspring?

*Note: The same letters are always used to depict the same gene for a trait in a cross, with the capital letter representing the dominant gene for the trait and the lowercase letter representing the recessive gene for the trait.

Genotype for a heterozygous tall pea plant = **Tt**

Genetype for a homozygous small pea plant = **tt**

T represents the tall gene

t represents the short gene

	T	t
t	Tt	tt
t	Tt	tt

Table 4.1

2 out of 4 offspring would be genotype **Tt** and

2 out of 4 offspring would be genotype **tt**

• • • • • •

Test Cross

An important type of cross conducted in many genetic experiments is called the **test cross**. The test cross does what its name implies, it is a test to determine an unknown genotype of an organism using a particular cross or mating.

Example

In guinea pigs, black hair is dominant to white hair. If a student needs to know if a guinea pig that he is studying is homozygous black or heterozygous black he would conduct a test cross. The genotype of the white guinea pig is known because a white guinea pig must have two genes for white hair, which is recessive. This fact makes crossing the black guinea pig with the white guinea pig the best way to figure out the genotype of the black pig. The Punnett squares show the two possibilities of these crosses.

B = black hair **b** = white hair

	B	b
b	Bb	bb
b	Bb	bb

	B	B
b	Bb	Bb
b	Bb	Bb

Table 4.2

In the Punnett squares in Table 4.2, the more black guinea pigs seen in the offspring, the more likely the parent black guinea pig was homozygous, or pure black.

• • • • • •

When someone tries to determine the genotype of an organism by mating an organism with an unknown genotype with a pure recessive organism, it is called a test cross.

Probability and Punnett Squares

Examples of Punnett squares were shown in Tables 4.1 and 4.2. It is important to remember that the results from Punnett squares are the probabilities of these events occurring. **Probability** is the study of the likelihood of the occurrence of random events. In studying these events, one is looking to be able to predict future events of the same type. In genetics, we look to predict the probability of certain genes from each parent ending up in the offspring. In probability, as the number of trials studied increases, the ratios obtained in the data will be closer to those predicted by the laws of probability or chance.

Example

When someone tosses a normal coin it has a 50% chance of landing on heads and a 50% chance of landing on tails. However, it is possible to toss a coin five times and get five heads. This does not mean the chance of getting heads is not 50%, because each time you toss that coin you still have a 50% chance of getting heads, or 50% chance of getting tails. If you were to toss the coin 1,000 times, the number of times the coin landed on heads would in most cases be much closer to 50% than if you conducted the experiment with only 20 tosses of this coin.

There is one chance to get *heads* on the coin (1). There is one chance to get *tails* on the coin (1). On one normal coin there are *heads* and *tails* (2). So the chance of getting either *heads* or *tails* is 1/2.

••••••

This same situation is true when using a Punnett square to predict the genetic outcome of offspring.

Example

If a hybrid tall pea plant is crossed with a hybrid tall pea plant, what is the probability the offspring will be tall?

T = tall gene   **t** = short gene

	T	**t**
T	**TT**	**Tt**
t	**Tt**	**tt**

The Punnett square shows the following predicted results

TT – tall pea plant
Tt – tall pea plant
Tt – tall pea plant
tt –short pea plant

} ¾ or 75% of the offspring are predicted to be tall

Table 4.3

••••••

If two tall hydrid plants are crossed, and you were to grow four seeds produced from this cross, the results shown in Table 4.3 do not mean definitely that three seeds will produce tall plants and one seed will produce a short plant. It is not impossible for all four seeds to grow into tall plants, or even for all four seeds to grow into short plants. All that probability is stating is that the results in Table 4.3 have the best chance of occurring. If the cross of these hybrid plants produced thousands of seeds, the chance that all of the seeds end up in the 3 (tall) to 1(short) ratio is very high.

Dihybrid Crosses

When Mendel crossed pea plants and looked at the way two traits were inherited by the offspring, the traits (genes) were distributed independently in all experiments conducted. You can use a Punnett square to show the possible offspring that could be produced from a cross of two organisms when you need to look at two traits at the same time. This type of cross is called a ***dihybrid cross***.

Example

In pea plants, the gene for tallness is dominant over the gene for shortness. In pea plants, the gene for yellow pea color is dominant over the gene for green pea color.

T = tall   **Y** = yellow seeds

t = short   **y** = green seeds

In a cross between these organisms, what could be the expected genotypes of the offspring?

TtYy × TtYy

To be able to complete the Punnett square, the possible genes that can be contributed from each parent offspring must be determined first so that they can be placed into the square.

Organism #1 **TtYy** can donate the following combinations to the offspring:

TY or **Ty** or **tY** or **ty**

The previous combinations show that the dominant gene for height could be donated with either the dominant gene for seed color or with the recessive gene for seed color. The recessive gene for height could be donated with the dominant gene for seed color or with the recessive gene for seed color.

Organism #2 can donate the following combinations to the offspring for the same reasons given above for organism #1:

TY or **Ty** or **tY** or **ty**

You then use these four combinations of possible genotypes from each organism to complete the Punnett square.

	TY	Ty	tY	ty
TY	**TTYY**	**TTYy**	**TtYY**	**TtYy**
Ty	**TTYy**	**TTyy**	**TtYy**	**Ttyy**
tY	**TtYY**	**TtYy**	**ttYY**	**ttYy**
ty	**TtYy**	**Ttyy**	**ttYy**	**ttyy**

Table 4.4

The offspring produced are:

9 = tall yellow plants   3 = tall green plants

3 = short yellow plants   1 = short green plants

It is important to note that any time a dihybrid cross is conducted between two heterozygous dominant plants, the expected ratio of the offspring will be in the ratio 9:3:3:1.

••••••

Example

In a cross between the following organisms, what could be the expected genotypes of the offspring?

To complete the Punnett square, the possible genes that can be contributed from each parent offspring must be determined first so that they can be placed into the square.

Organism #1 **TtYy** can donate the following combinations to the offspring:

TY or **Ty** or **tY** or **ty**

The above combinations show that the dominant gene for height could be donated with either the dominant gene for seed color or with the recessive gene for seed color. The combinations also show that the recessive gene for height could be donated with the dominant gene for seed color or with the recessive gene for seed color.

Organism #2 **ttyy** can donate the following combinations to the offspring:

ty or **ty** or **ty** or **ty**

The previous show that the recessive gene for height could be donated to either recessive gene for seed color and that the recessive gene for seed color could be donated with either recessive gene present for height.

You then use these four combinations of possible genotypes from each organism to complete the Punnett square shown here:

T = tall   **t** = short   **Y** = yellow   **y** = green

	TY	TY	tY	ty
ty	TtYy	Ttyy	ttYy	ttyy
ty	TtYy	Ttyy	ttYy	ttyy
ty	TtYy	Ttyy	ttYy	ttyy
ty	TtYy	Ttyy	ttYy	ttyy

Table 4.5

The offspring produced are:

4 = tall yellow plants 4 = tall green plants

4 = short yellow plants 4 = short green plants

••••••

The results of dihybrid crosses like the ones shown previously were used to verify the Law of Independent Assortment, because these crosses show that all the genes are inherited independently.

Some organisms have traits that are neither dominant nor recessive. Organisms with these traits, which have a heterozygous genotype, show a form of blending of the two traits in question. Two such examples of this genetic condition are codominance and incomplete dominance.

Codominance

In shorthorn cattle, pure red hair and pure white hair genotypes are possible. Both the red hair and the white hair trait in these cattle are considered to be dominant. The offspring produced from a cross between a pure red hair and pure white hair cattle is called *roan*. Roan means that the cattle have some hair that is white and some hair that is red. In **codominance**, both dominant traits are expressed independently in the phenotype.

Example

R = pure red hair **W** = pure white hair **RW** = roan hair color

	W	W
R	RW	RW
R	RW	RW

Table 4.6

All the offspring will contain the dominant trait for white hair and the dominant trait for red hair and when expressed together these cattle will have the genotype **RW** (roan hair color). Roan hair color is produced from separate white and separate red hair growing together on the body of the cattle.

••••••

Incomplete Dominance

In a plant called the snapdragon there are red flowers and white flowers possible. When a pure red plant is crossed with a pure white plant, the offspring will produce pink flowers. In **incomplete dominance** there is a blending effect seen in the phenotype of the hybrid genotype. In this type of dominance each gene is partially expressed.

Example

In a cross between two flowers, one with pure red petals and one with pure white petals, the following results occur:

R = red flower color **W** = white flower color **RW** = pink flower color

	R	R
W	RW	RW
W	RW	RW

Table 4.7

All the offspring will contain the dominant trait for white petals and the dominant trait for red petals, and when expressed together these flowers will not have any white or red flowers but a blending of the two traits, producing, in this case, pink (**RW**). Pink flower petals are produced from the genotype of red and white coming together in the condition called incomplete dominance.

••••••

Multifactorial Inheritance

In Mendel's work with pea plants, the phenotypes expressed in these plants were all shown to be produced by the action of two genes. But two or more genes control some traits of organisms. Multifactorial inheritance is the condition when two or more genes control one trait. Because usually more than two genes control the outcome of the trait, the characteristics expressed in the phenotype of these genes produce a wide range of results. An example of this condition would be skin color in humans. Skin color is controlled by at least four genes, and because of this fact many color tones of the skin can be produced.

Multiple Allelism

For each gene present in an organism, there are usually two possible forms of that gene called alleles. These two genes usually are in the forms dominant and recessive. But for some genes there will be three or more alleles present for a particular trait. This does not mean one individual organism will have more than two alleles for any one gene, but that more than two alleles are present for the one gene within the population of the species. For example, in humans there are three alleles possible for blood type—A, B, or O—but each human has, at most, two of those alleles possible.

Practice Section 4-1

1. The phenotype of a guinea pig's hair color can be best determined by
   a) test cross c) pedigree chart
   b) looking at the guinea pig d) knowing the genotype of the mother pig
2. If a trait that is visible in the parent organisms is not seen in the offspring, but then returns in the F2 generation, the most probable cause is that this trait in question is
   a) recessive b) dominant c) codominant d) mutated
3. Which of the following genotypes below shows a pure dominant genotype?
   a) Aa b) aa c) AA d) AB
4. Genes that are located at the same position on homologous chromosomes are called
   a) recessive b) dominant c) alleles d) albinism
5. In garden pea plants, the offspring formed from a cross between two heterozygous tall parents would most likely be
   a) 25% tall b) 50% tall c) 75% tall d) 100% tall
6. A pink flower is produced by crossing a plant that has white flowers with a plant that has red flowers. This example most likely shows what condition in genetics?
   a) incomplete dominance c) dominance
   b) codominance d) recessive trait
7. Who was Gregor Mendel?
8. What is the difference between a dominant trait (allele) and a recessive trait (allele)?
9. What is the genetic idea of *incomplete dominance*?

Lesson 4-2: The Work of Thomas H. Morgan

Thomas H. Morgan made great advances in the field of genetics using as a test organism the fruit fly, ***Drosophila melanogaster***. The fruit fly turned out to be an excellent organism in which to study the patterns of genetics because:

1. Fruit flies produce many offspring.
2. Fruit flies are very small in size.

3. Fruit flies have only eight chromosomes.
4. Fruit flies have a short life cycle.
5. Fruit flies have very large chromosomes.

All the reasons listed here made caring for, studying, and obtaining accurate scientific data from the genetics of the fruit fly much easier than it had been for previously studied organisms. While studying fruit flies, Morgan and his associates also discovered how the sex of the offspring is determined. The X chromosome and the Y chromosome are called the sex chromosomes in the organism. The sperm can carry either an X or a Y chromosome. The egg always contains an X chromosome.

The Punnett square shows how the sex of the offspring is determined with the union of egg and sperm.

	X	Y
X	XX	XY
X	XX	XY

Table 4.8

A fertilized egg with the XX chromosome will become a female. And a fertilized egg with the XY chromosome will become a male.

Linkage

It was originally thought by Mendel and stated in the Law of Independent Assortment that each trait of an organism is inherited independently from every other trait. This means that when a gene is donated to the offspring from one of the parents, that gene is donated alone. Through the years, as more genetic experiments were conducted, exceptions to this rule were observed. Some traits always seemed to be present when other traits were also present in the organism. The presence of one trait with another occurs because genes on the chromosomes sometimes stick together, or "link." This means **linkage** is a condition where one of the genes donated by the parent organism has another gene attached to it from the same chromosome.

In the late 1930s, the gene theory was developed in large part because of the work done by Thomas Morgan.

The gene theory includes the following ideas:

1. Chromosomes are made up of genes that are arranged in a row. A beaded necklace is a good analogy of this concept.
2. Genes can become linked on the same chromosome, which means that some traits observed can, at times, be inherited together. An example of this in humans is red hair and freckles.

3. Crossing over, at times, breaks these linked genes so that they are not inherited together.
4. Gene maps can be created by the position of genes on the chromosome.

••••••

Studying the frequency at which linked genes on a chromosome separate and then eventually recombine was used to create a gene map of the distances between genes on the chromosome. This information about gene mapping was used to eventually put together the human genome map.

Sex Linkage

The work on the fruit fly also made clear a situation that had been noticed for many years but was not fully explainable. Many more men are color-blind than woman. More men than women have hemophilia, a condition where the blood clots very slowly. When mutations in the fruit fly were studied closely, it was noticed that some traits were acquired in much greater frequencies by the male flies compared to the female flies. It was discovered that these conditions were caused by problems on the sex chromosomes of the organism. If you recall from earlier in this chapter, we discussed that male organisms have an X and a Y chromosome, while female organisms have two X chromosomes. It turns out that some genes important for chemical processes within the body reside on the X chromosome. For these processes to work correctly, the organism needs at least one good copy of the gene from the X chromosome. Because males have only one X chromosome, they have only one shot to get the good gene; but females have two X chromosomes, and so have two shots to get one good gene. This is why these conditions are more common in males. In the Punnett square in Table 4.9, you can see mathematically why the chances of a male having a sex-linked condition are better than a female's.

Example

A normal human male has a child with a female with one defective X chromosome gene. This woman would not show any signs of the condition, but sould be able to pass on the bad gene to her offspring. This type of genotype condition is called a **carrier**.

X: normal X sex chromosome **X–**: defective X sex chromosome

	X	**X–**
X	**XX**	**XX–**
Y	**XY**	**X–Y**

Table 4.9

You can see from the example that 50% of the male offspring will have the genetic disease from this cross and none of the female offspring will have the disease, although 50% of the female offspring will be carriers.

••••••

The work of Thomas H. Morgan and the later discovery of the structure of the DNA molecule have greatly advanced our knowledge of the gene.

Table 4.10 is a chart of some of the more common sex-linked diseases present in humans.

Sex-Linked Diseases

Disease	Condition
Hemophilia	X-linked recessive disorder that when present does not allow the blood to clot properly.
Color blindness	X-linked recessive disorder.
Klinefelter syndrome (XXY)	Double X disorder that causes behavioral and speech problems. Person affected is usually tall and infertile.
Turner's syndrome (XO)	Only one X chromosome that can lead to underdeveloped ovaries, abnormal jaws, webbed neck, and flat chest.
(XYY)	Double Y disorder that has been shown to cause a greater instance of aggression in affected individuals.

Table 4.10

Practice Section 4-2

1. What genetic condition probably explains the fact that most people who have red hair also have freckles?

   a) codominance c) mutation
   b) incomplete dominance d) linkage

2. The parents of three girls are expecting another child, what are the chances the child will be a boy?

   a) 25% b) 50% c) 75% d) 100%

3. Linked genes can be separated on a chromosome by the process of

   a) crossing over b) mutation c) linkage d) nondisjunction

4. If a color-blind man and a woman who is a carrier for color blindness have a child, what percentage of their female children will be color-blind?

   a) 0% b) 25% c) 50% d) 100%

5. Which of the following conditions would not be desirable in a genetic study?
   a) Testing organisms that produce large numbers of offspring.
   b) Testing organisms that are small in body size.
   c) Testing organisms that have a long life cycle.
   d) Testing organisms that have large chromosomes.
6. Which of the following would represent the sex chromosomes of a normal female?
   a) XX b) XY c) YY d) XXY
7. What is meant by the genetic term "linkage"?
8. What is a carrier?
9. Why is genetic color blindness considered to be a sex-linked disease? Why are males much more likely to be color blind then females?

Lesson 4-3: Molecular Genetics

Molecular genetics is the study of the chemistry of the materials of inheritance. This study focuses in on the structure of **DNA** and its ability to organize the construction of all the molecules needed for life.

The Structure of DNA

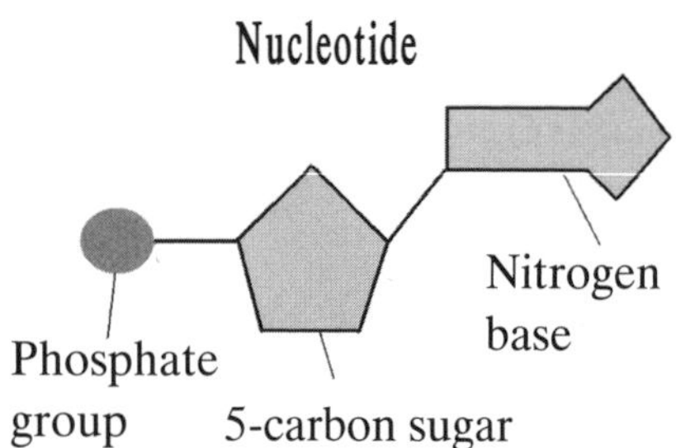

Knowledge of the structure and function of DNA is imperative for an understanding of the chemistry of inheritance. DNA, stands for ***deoxyribose nucleic acid,*** and is the chemical polymer that is the structure and function of inheritance. A DNA molecule is made up of thousands of repeating smaller parts called nucleotides. A **nucleotide** consists of three parts: a deoxyribose molecule, a phosphate molecule, and a nitrogen base molecule.

There are four nitrogen bases that make up the "rungs" of the DNA molecule, they are cytosine, guanine, thymine, and adenine.

These bases are connected at that middle of the DNA molecule by weak hydrogen bonds. The bases are bonded at the middle of the DNA molecule in what is called ***complementary pairs.*** This means that each nitrogen base will only bond with another specific nitrogen base. The nitrogen base adenine will only bond with thymine, and cytosine will only bond with guanine in a DNA molecule.

The specific nature in which these nitrogen bases are bonded becomes very important in the creation of the code of life that we will discuss later in this chapter.

Reading the DNA

Reading a DNA molecule centers in on the nitrogen bases present in the molecule, and their arrangement. A row of three nitrogen bases forms what is called a ***triplet*** and is called a **codon**. Codons can be looked at as the words of the DNA language. The number of codons and their arrangement in a particular DNA molecule creates specific instructions that other molecules will read to form the structures of life. For example, the codon CCG is the triplet code for making the amino acid ***glycine.*** Various triplets of nitrogen bases make all of the 20 essential amino acids of life. These various amino acids are then bonded in different arrangements to create the many proteins necessary for the structures of life.

DNA Replication

In the nuclei of most cells you will find the chemical of inheritance DNA. The DNA is a very precious substance that must be protected, for it controls both the present and the future life of the cell. For this reason, in most cases DNA will not leave the protection of the cell's nucleus. The problem is that for the cell to construct the structures of life and carry on the many chemical processes of life, it is required that the organelles in the cytoplasm use the DNA. Cells have found an ingenious way of bringing the "commands" of the DNA to the cell's cytoplasm and beyond, while keeping the DNA within the nucleus.

DNA Molecule Unzipping

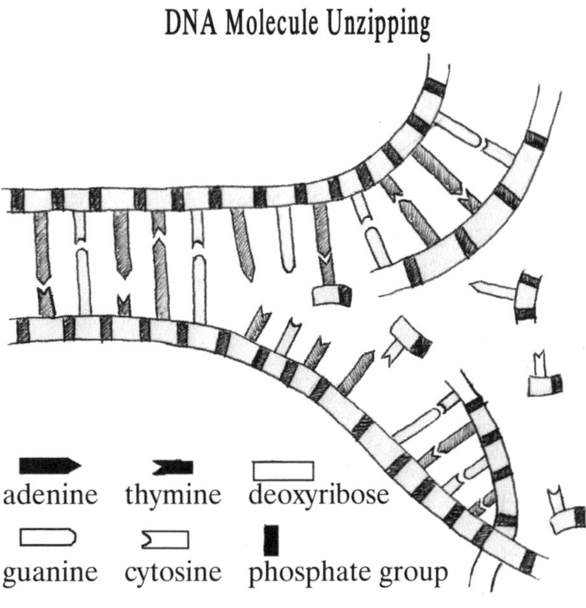

A special chemical called **mRNA** (messenger ribose nucleic acid) makes a copy of the DNA code and brings this copy to the necessary structures within the cell's cytoplasm. For replication to occur, the DNA molecule must "unzip" at the hydrogen bonds that hold the nitrogen bases together. Each piece of the DNA molecule is called a ***template*** and will be used to create complementary copies of each strand.

As the DNA molecule separates at the center, an enzyme known as ***DNA polymerase*** begins to build a chain of complementary bases along the exposed nitrogen bases of each one of the DNA strands. The RNA strand being created in this process will have all the complementary bases of the original DNA strand with one exception. The base ***uracil*** is used as a complement to adenine instead of the usual base thymine. The presence of uracil in a molecule is usually a good indicator that the molecule in question is mRNA.

Example

If a DNA strand has the bases C-C-G-A-T-G, then the strand reproduced by the mRNA will have the complementary bases G-G-C-U-A-C.

••••••

This biological process of mRNA copying the original DNA molecule is called **transcription**.

Making Proteins: Translation

The chemical process of **translation** creates the proteins needed by the living organism. After the process of transcription has finished in the nucleus of the cell, the mRNA with the complementary base code from the original DNA molecule makes its way out of the nucleus and into the cytoplasm of the cell. In the cytoplasm of the cell, the mRNA travels next to the ribosomes. It is because of this journey of bringing the code of the DNA to the ribosomes that this particular RNA is called the messenger RNA. The ribosomes act as a surface area for the bonding of long chains of amino acids based on the code present on the mRNA. A second type of RNA called **tRNA** (transfer RNA) brings these amino acids to the mRNA and ribosomes. The tRNA molecule is specifically constructed to carry amino acid molecules from the cytoplasm of the cell to these bonding sites in the ribosomes. The type of amino acid that each tRNA molecule carries is determined by a sequence of three nitrogen bases located at the other end of the molecule. These three nitrogen bases, which are complementary to the bases on the mRNA, are called the ***anticodon***.

Transfer RNA (tRNA)

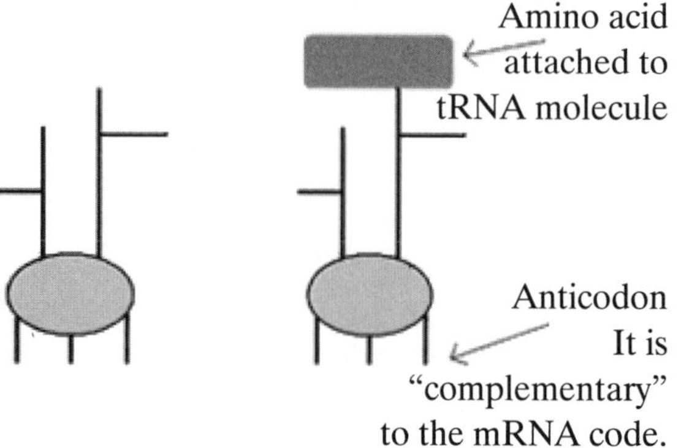

Imagine, for example, that an mRNA molecule has a portion of its surface that contains the bases G-G-U. This portion of the mRNA will bind with a section of the tRNA anticodon with the nitrogen bases in the sequence C-C-A. As many tRNA molecules come into the ribosome, with amino acids attached at one end, their anticodon on the other end will direct them where to engage the mRNA. The code on the mRNA aligns the tRNA molecules up in the correct order, which in turn aligns the amino acids they are carrying in the correct order to make specific proteins. The tRNA molecules only remain for a moment at the bonding site of the ribosomes until the amino acids they deliver bond together to form the correct protein.

The chemical process of translation, illustrated here, requires the action of many enzymes working in many locations throughout the surface of the ribosome.

Translation (Occuring in the Ribosome of the Cell)

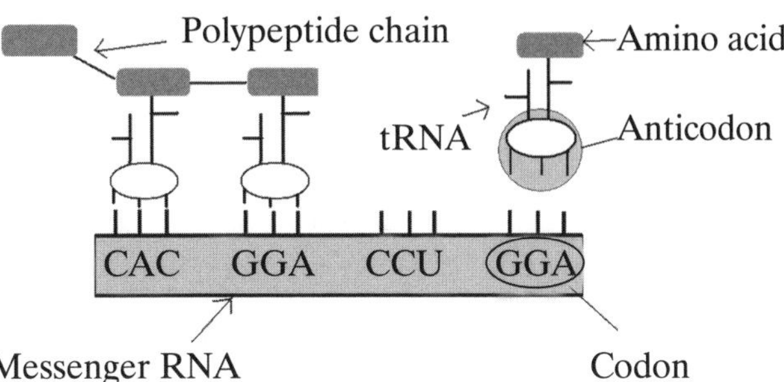

Practice Section 4-3

1. Which of the following is NOT part of a DNA molecule?

   a) nitrogen base c) phosphate group
   b) deoxyribose sugar d) ribose sugar

2. What would the complementary RNA sequence be for the base sequence on part of the DNA molecule shown here?

   A – G – G – C – T

   a) A – G – G – C – T c) U – C – C – G – A
   b) T – C – C – G – A d) G – A – A – T – C

3. The language of the genetic material is written in three nitrogen base groups called

   a) thymines b) codons c) nondisjunctions d) deletions

4. A sequence of DNA that has 150 nucleotides would produce a polypeptide chain

   a) 150 amino acids long c) 50 amino acids long
   b) 300 amino acids long d) 450 amino acids long

5. The sequence of nucleotides in the messenger RNA molecule is determined by the sequence of nucleotides in

   a) transfer RNA b) protein c) DNA d) amino acids

6. What molecule brings amino acids to the ribosomes in the cytoplasm during protein synthesis?
   a) mRNA b) tRNA c) DNA d) PGAL
7. What is the structural difference between a molecule of DNA and a molecule of RNA?
8. Explain the genetic process of translation.
9. What is a codon?

Lesson 4-4: Gene and Chromosomal Mutations

Mutations are sudden changes in the chemical make-up of the DNA that controls the chemical actions of the gene or chromosome. These changes in the DNA can cause the creation of abnormal proteins. Many times these abnormal proteins formed by mutations are harmful or fatal to the organism. Mutations are caused when the cell copies its DNA incorrectly. The rate of mutations occurring in the DNA of cells is very low, but certain environmental conditions such as radiation and certain specific chemicals can increase the rate of mutations. Mutations can affect the DNA in body or somatic cells; it can also affect the DNA in germ or sex cells. If the mutation is present in the body cells, only that particular organism will be affected by the results of the mutation. If the mutation is present in the sex cells, the parent organism and its offspring can be affected by the change in DNA.

Mutations that affect only a single gene are called ***gene mutations***. A gene mutation is when a section of the DNA that makes up a gene is altered. Gene mutations, although rare, can occur during normal cell division, but the following conditions are thought to increase the rate of gene mutations:

- Many forms of radiation such as X rays, cosmic rays, and ultraviolet rays.
- Some chemicals such as PCBs, saccharin, and nitrites.

The genes that make up a chromosome are created by nucleotide sequences or codons. The nitrogen bases that make up a codon can be altered during a mutation. When a nitrogen base is altered, the codon is changed, and this will mean that the protein it was suppose to make will be changed.

The three ways a sequence of nitrogen bases can be affected by a mutation are:

1. **Substitution:** A condition where one nitrogen base is replaced by another nitrogen base in a codon.
2. **Insertion:** A condition where a nitrogen base is "inserted" into a previously existing codon.
3. **Deletion:** A condition where a nitrogen base is removed or "deleted" from a previously existing codon.

Insertion and deletion are sometimes referred to as **frameshift mutations**, because the addition or deletion of a single nitrogen base in a string of nitrogen bases will shift how all of these bases will be read in their now changed groups of codon sequences.

A **chromosome mutation** is when the number of chromosomes present in the organism changes or the structure of one particular chromosome changes.

The two types of chromosome mutations are:
nondisjunction and ploidy.

Nondisjunction is when only one chromosome goes in the wrong direction during anaphase of cell division. This creates a cell that has one extra chromosome. ***Down syndrome*** is a chromosome mutation where the individual has one extra number 21 chromosome.

Ploidy is when all the chromosomes end up on one side of the cell at the end of anaphase of cell division. This could mean a sperm with a diploid number of chromosomes, instead of monoploid, fertilizes an egg that is also diploid. The offspring produced will then be 4n (tetraploid).

Practice Section 4-4

1. A mutation is when a cell

   a) does not accurately copy its DNA c) ingests food

   b) reproduces d) produces waste products

2. A condition when homologous chromosomes fail to separate during cell division is known as

   a) polyploidy b) nondisjunction c) disjunction d) cleavage

3. Which of the following could cause a gene mutation?

   a) gamma radiation c) nitrites

   b) UV radiation d) all of the above

4. Looking at the codon sequences here, how would this mutation be classified?

   ATC – TGC – GGA DNA sequence before mutation

   ATT – CTG – CGG DNA sequence after mutation

   a) insertion b) deletion c) substitution d) none of these

5. A man develops a mutation of cells in his skin from exposure to the sun, which of the following is not likely to occur?

   a) The mutation could lead to skin cell cancer

   b) The mutation could interfere with the skin cells ability to protect the body

c) The mutation could be passed on to his children

d) The mutation could lead to changes in the structure of the skin cells.

6. Down syndrome results from a genetic mutation called

   a) histamine b) nondisjunction c) insertion d) substitution

7. What are some factors that can cause gene mutation?
8. Explain the difference between an *insertion* and a *deletion*?
9. What type of gene mutation causes the genetic condition of Down syndrome? How does it occur?

Terms From Chapter 4		
asexual reproduction	genotype	phenotype
carrier	heredity	Punnett square
codominance	heterozygous	recessive trait
codon	homozygous	sexual reproduction
cross-pollination	incomplete dominance linkage	test cross
DNA	mRNA	transcription
dominant trait	mutation	translation
genetics	nucleotide	tRNA

Chapter 4: Exam

Matching Column for Genetic Crosses

Match the genetic cross of the parents on the left with the genotypes on the right of the offspring most likely to be produced from that cross. You may use an answer choice more than once or not at all.

Genetic Cross	Predicted Offspring
1. BB × bb	a. 25% BB 50% Bb 25% bb
2. Bb × Bb	b. 100% bb
3. BB × BB	c. 100% Bb
4. bb × bb	d. 50% Bb 50% bb
5. Bb × bb	e. 100% BB
	f. 75% Bb 25% bb

Matching Column for Genetic Crosses

Match the genetic cross of the parents on the left with the phenotypes on the right of the offspring most likely to be produced from that cross.

You may use an answer choice more than once or not at all.

(T= tall; t =short)

Genetic Cross	Predicted Offspring
6. TT × tt	a. 100% short
7. tt × Tt	b. 75% tall 25% short
8. Tt × Tt	c. 100% tall
9. TT × TT	d. 25% tall 75% short
10. tt × tt	e. 50% tall 50% short

Multiple Choice

11. What organisms did Mendel use in his famous genetic experiments?
    a) mice b) pea plants c) rose plants d) horses
12. Which of the following shows a pure dominant genotype?
    a) Tt b) tt c) TT d) none of these
13. What would be the expected phenotypes of the offspring from the following cross?

    TT × Tt

    a) 50% tall, 50% short c) 100% short
    b) 100% tall d) 75% tall, 25% short
14. What is the name given to a particular cross used to determine the unknown genotype of an organism?
    a) test cross c) heterozygous cross
    b) homologous cross d) target cross
15. In humans, skin color is controlled by at least four genes, this is an example of
    a) multiple allelism c) linkage
    b) multifactorial inheritance d) nondisjunction
16. The fruit fly turned out to be a very good organism for genetic studies because
    a) it produces many offspring c) it has very large chromosomes
    b) it has a short life cycle d) all of these
17. A female that has the genotype for color blindness but not the phenotype for color blindness would be considered
    a) color-blind
    b) a carrier for color blindness
    c) homozygous dominant for color blindness
    d) homozygous recessive for color blindness

18. If a human male that is homozygous for blood type B has a child with a woman who is heterozygous for blood type A, what percent of their children can be expected to have blood type B?
    a) 25% b) 50% c) 75% d)100%
19. A molecule of DNA could be composed of all of the following except
    a) deoxyribose sugar c) phosphate group
    b) uracil d) cytosine
20. Three nitrogen bases form a triplet, which is sometimes called a
    a) codon b) ribosome c) nucleotide d) peptide
21. The biological process of mRNA copying the original DNA molecules is called
    a) translation c) dehydration synthesis
    b) transcription d) hydrolysis
22. How many nitrogen bases are needed to code for one amino acid?
    a) 1 b) 2 c) 3 d) 4

Short Response

23. Describe the differences between DNA replication and RNA transcription.
24. A female that is a carrier for color blindness has a child with a man that is color-blind. If the child is a boy, what are the chances that he will be color-blind?
25. If a woman had a child after one of her body cells experienced a genetic mutation, what effect would this have on the child? If a woman had a child after one of her sex cells experienced a mutation, what effect would this have on the child?
26. In the garden pea plants studied by Mendel, the pea came in two colors—yellow and green. In humans, skin color comes in many different shades. Explain why there is this difference between the skin color in humans and the color in peas.

Answer Key

Answers Explained Section 4-1

1. **B** is the correct answer because the phenotype of a guinea pig can best be determined by looking at the characteristics of that organism.

   A is not correct because test cross is used to determine the unknown genotype of an organism by mating it with a homozygous recessive organism.

   C is not correct because a pedigree chart is a family tree of the known genotypes of members of that family.

   D is not correct because knowing the genotype of the mother pig does not necessarily mean you will be able to predict the phenotype of the offspring.

2. **A** is the correct answer because a recessive trait can disappear phenotypically and reappear later in the next generation, because in order for a recessive gene to be expressed phenotypically, it must be matched with another recessive gene of the same characteristic.

   B is not correct because dominant traits will always be expressed phenotypically.

   C is not correct because codominant traits are two dominant traits that are expressed together phenotypically.

   D is not correct because mutated is when a gene or chromosome is chemically altered.

3. **C** is the correct answer because AA is the correct symbols for a pure, dominant genotype.

   A is not correct because Aa is the symbol for a heterozygous genotype.

   B is not correct because aa is the symbol for a homozygous recessive genotype.

   D is not correct because AB is the symbols for a codominant genotype.

4. **C** is the correct answer because genes located at the same position on homologous chromosomes are called alleles.

   A is not correct because a recessive gene is a gene that is only expressed phenotypically when matched with other recessive gene.

   B is not correct because a dominant gene is a gene that is always expressed phenotypically when present.

   D is not correct because albinism is the genetic condition where an organism does not produce the needed pigments of the skin, eyes, and hair.

5. **C** is the correct answer because a cross between Tt × Tt would produce 75% (T) tall plants. The other 25% of the offspring can be predicted to be (t) short plants.

6. **A** is the correct answer because the production of a pink flower from a cross between a plant with red flowers and a plant with white flowesr is an example of incomplete dominance. Incomplete dominance is when both dominant traits are only partially expressed.

   B is not correct because codominance is when two dominant traits are both expressed fully phenotypically at the same type.

   C is not correct because dominance is when one gene is phenotypically expressed over another gene.

   D is not correct because a recessive trait is only expressed when two recessive genes for that particular trait are matched together in the genotype.

7. Gregor Mendel is considered to be the father of Genetics. Mendel became famous because of his study of the inheritance of certain traits in pea plants. Mendel showed that the inheritance of these traits follows particular laws, which were later named after him. The significance of Mendel's work in the field of genetics was not recognized until the 20th century.

8. A dominant allele is an allele that will be expressed when paired with a recessive allele. A dominant allele is what creates a dominant trait. Brown eyes are a dominant trait in humans, caused by the dominant allele for brown eyes. Blue eyes are a recessive trait in humans, caused by the recessive allele for blue eyes. If a brown eye allele is matched with a blue eye allele, the brown eye trait will only be expressed in the individual because brown is dominant and blue is recessive.

9. Incomplete dominance is a form of intermediate inheritance in which one allele for a specific trait is not completely dominant over the other allele. This results in a combined phenotype. For example if an allele for red color and a allele for white color in flower petals are combined, because neither color allele is completely dominant over the other color, the offspring plants will have pink colored petals.

Answers Explained Section 4-2

1. **D** is the correct answer because the genetic condition that results in some traits like freckles and red hair usually being inherited together is linkage.

   A is not correct because codominance is when two dominant traits are fully expressed together.

   B is not correct because incomplete dominance is when two dominant traits are each partially expressed together.

   C is not correct because a mutation is when a gene or chromosome is chemically altered.

2. **B** is the correct answer because there is a 50% chance of having a boy. No matter how many female or male children this couple has had, there will always be a 50% chance of having a boy and a 50% chance of having a girl.
3. **A** is the correct answer because linked genes can be separated by the process of crossing over. Crossing over is when two homologous chromosomes exchange genes.

   B is not correct because mutation is when a gene or chromosome is chemically altered.

   C is not correct because linkage is when genes on the same chromosome stay together and are then inherited together.

   D is not correct because nondisjunction is when homologous chromosomes fail to separate during meiosis.
4. **C** is the correct answer because there is a 50% chance of a having a color-blind female child when a color-blind man and a woman who is a carrier for color blindness have a child. This 50% probability is determined by a Punnett square mating of carrier female X–X with a color-blind male X–Y.
5. **C** is the correct answer because testing organisms that have a long life cycle is not desirable because it takes too long to complete generational studies of these organisms.

   A is not correct because large numbers of offspring are desirable because it allows for many test subjects.

   B is not correct because testing organisms that are small in body size is desirable because it allows for many organisms to be held in a small area.

   D is not correct because testing organisms that have large chromosomes is desirable because large chromosomes are easier to observe and study.
6. **A** is the correct answer because the sex chromosomes of a normal female would be represented by XX.

   B is not correct because XY represents the normal sex chromosomes of a male.

   C is not correct because YY is not a possible arrangement of sex chromosomes.

   D is not correct because XXY is a mutation of the sex chromosomes called Klinefelter Syndrome.
7. Genetic linkage is a term which describes the tendency of certain alleles to be inherited together. Alleles on the same chromosome are physically close to one another and tend to stay together during meiosis, and are then genetically linked. This means that the appearance of one trait in an individual is usually

connected with another different trait because they have been brought together to the individual during linkage. An example of this occurrence in humans is red hair and freckles which are usually linked and obtained together.

8. A carrier is an organism that has inherited a genetic trait but does not display that trait. An organism that is a carrier is able to pass the gene to their offspring, who may then be able express the gene if the individual is homozygous recessive for that passed gene. For example, the gene that causes the disease Cystic Fibrosis is a recessive gene that can be carried by an individual who shows no signs of the disease. If two individuals, who are both carriers of this disease gene, have offspring, there is a 25% chance their child could acquire a homozygous pair of Cystic Fibrosis alleles, a gene from each parent. This child would not be a carrier but would actually have the disease, even though both parents would be perfectly healthy.
9. Many of the genes associated with color vision are found on the X sex chromosome. This means that during the fertilization process the genes that are acquired on the X chromosome will help determine the individuals ability to see color. Females have two X sex chromosomes, while males have only one X sex chromosome. This is why color blindness is much more common in males. Males have only one shot to get the correct genes for color vision, while females have two shots to get the correct genes.

Answers Explained Section 4-3

1. **D** is the correct answer because ribose sugar is not part of a DNA molecule. Ribose sugar is part of RNA.

   A, **B**, and **C** are not correct because nitrogen bases, deoxyribose sugar, and a phosphate group are all parts of a DNA molecule.

2. **C** is correct because U-C-C-G-A is the complementary RNA strand. In the complementary RNA molecule the following nitrogen bases are matches for the nitrogen bases found in the DNA molecule:

DNA Template Molecule	RNA Complementary Strand
Adenine	Uracil
Thymine	Adenine
Guanine	Ctosine
Cytosine	Guanine

Table 4.11

3. **B** is the correct answer because codon is the name of three nitrogen bases taken together in the genetic language.

A is not correct because thymine is a nitrogen base found in DNA molecules.

C is not correct because nondisjunction is when homologous chromosomes fail to separate.

D is not correct because deletion is a chromosomal mutation where a nitrogen base is removed from the original codon sequence.

4. **C** is the correct answer because 150 nucleotides would produce a polypeptide chain 50 amino acids long, because for each amino acid produced you must have a group of three nucleotides.

5. **C** is the correct answer because the DNA determines the nucleotide sequence in the messenger RNA.

   A is not correct because transfer RNA brings amino acids to the ribosomes during protein synthesis.

   B is not correct because proteins are organic compounds created from chains of subunits called amino acids.

   D is not correct because amino acids are the building blocks of protein. Amino acids are brought to and assembled at the ribosome by the transfer RNA.

6. **B** is the correct answer because the tRNA (transfer RNA) brings the amino acids to the ribosomes for protein synthesis.

   A is not correct because mRNA makes a complementary copy of the DNA from the nucleus and brings this information to the cytoplasm.

   C is not correct because DNA is the blueprint of all genetic codes in the cell.

   D is not correct because PGAL is the first organic compound produced in photosynthesis.

7. DNA (Deoxyribose nucleic acid) is a nucleic acid molecule constructed of a deoxyribose sugar molecule while RNA (ribose nucleic acid) is a nucleic acid molecule constructed of a ribose sugar molecule. Ribose is a water soluble, pentose sugar. Deoxyribose is a ribose sugar molecule that has lost an oxygen molecule.

8. In translation, protein molecules are created by the connection of messenger RNA with transfer RNA. Transfer RNA brings the correct amino acids to the messenger RNA strand that will attach the amino acids together in the correct sequence to create the needed protein molecule. Enzymes enable this process to happen very quickly and efficiently.

9. The genes that code for particular proteins are composed of three nucleotide units called codons. An example of a particular codon would be something like CUG. (C – cytosine, U – uracil, and G – guanine) This particular codon, CUG, is used to create the amino acid called Leucine.

Answers Explained Section 4-4

1. **A** is the correct answer because a genetic mutation is when a cell does not accurately copy its DNA.

   B is not correct because when a body cell reproduces, it is called mitosis, and when a sex cell reproduces it is called meiosis.

   C is not correct because when a cell ingests food through its cell membrane, those are the life processes of nutrition and transport.

   D is not correct because when a cell removes waste products, it is the life process of excretion.

2. **B** is the correct answer because when homologous chromosomes fail to separate during cell division, one cell will get too many chromosomes and one cell will not get enough chromosomes, this is a condition called nondisjunction.

   A is not correct because polyploidy is a gene mutuation where an organism has an extra set of chromosomes.

   C is not correct because disjunction is not a biological term for cell mutations.

   D is not correct because cleavage is when a fertilized cell begins to divide by mitosis into more cells.

3. **D** is the correct answer because gamma radiation, UV radiation, and nitrites have all been linked to genetic mutations in organisms.

4. **A** is the correct answer, which is insertion, because in the first codon the nitrogen base thymine has been inserted at the end. This condition has caused all the other nitrogen bases after that to shift to the right.

   B is not correct because deletion is a mutation when a nitrogen base is removed from a codon in a genetic sequence.

   C is not correct because substitution is a mutation where one nitrogen base is replaced by another nitrogen base in the codon.

5. **C** is the correct answer because mutation of body or somatic skin cells will not be passed on to the next generation. Usually if the mutation occurs in the sex cells of the organisms then the mutation will be passed on to the next generation.

   A is not correct because mutation of skin cells from radiation can lead to the abnormal division of skin cells or cancer.

   B is not correct because any type of mutation to a cell can alter its ability to perform certain functions, such as protection.

D is not correct because mutations can lead to abnormal cell division, which can then lead to changes in the structure of the cells making up to the skin.

6. **B** is the correct answer because the genetic condition called Down's syndrome results from the mutation called nondisjunction. Individuals with Down's syndrome, because of nondisjunction, have one extra 21st chromosome.

   A is not correct because histamine is not a genetic mutation, it is a chemical produced in the body that causes the inflammation of tissue.

   C is not correct because insertion is a genetic mutation where a nitrogen base is added to a codon, which then shifts all following nitrogen bases.

   D is not correct because substitution is a genetic mutuation where a nitrogen base is substituted to another nitrogen base in a particular codon.

7. Gene mutations can be caused by radiation and chemicals like PCB's and nitrites. Even some viruses can cause gene mutations.
8. An insertion is when a nitrogen base, like adenine or guanine, is inserted into a previously existing codon of bases. This insertion changes the pattern of all the bases that come after it. A deletion is when a nitrogen base is removed or deleted from a previously existing codon of bases. This removal of a base also has the harmful affect of changing all the codon patterns that follow it.
9. Down syndrome is a chromosome mutation where an individual is born with one extra 21st chromosome. The chromosome mutation of *nondisjunction* is when one chromosome goes in the wrong direction during cell division causes one of the sex cells to have an extra chromosome.

Answers Explained Chapter 4 Exam

Matching Column

1. C
2. A
3. E
4. B
5. D
6. C
7. E
8. B
9. C
10. A

Multiple Choice

11. **B** is correct because Mendel used pea plants in his famous genetic experiments conducted in the gardens of his monastery.
12. **C** is correct because a capital letter represents a dominant trait and two capital letters together would mean the organism is pure for that trait (TT).

A is not correct because the genotype Tt is heterozygous, not pure, for the trait in question.

B is not correct because the genotype tt is homozygous, or pure, but recessive not dominant for the trait in question.

13. **B** is correct because in the cross between TT and Tt using a Punnett square every offspring will contain at least one T gene, which means all 100% of the offspring will appear tall.

    A, **C**, or **D** are not correct because for any of these choices to be right the recessive gene would have to be alone in one of the possible offspring genetic combinations produced from the Punnett square, and this does not happen.

14. **A** is the correct answer because a test cross is when a recessive homozygous organism is mated with an organism with an unknown genotype to determine the genotype of that unknown organism.

    B, **C**, and **D** are not crosses done in genetics.

15. **B** is the correct answer because multifactorial inheritance is when two or more genes control the expression of one trait. The skin tones of human beings can be a wide range of colors because of this condition.

    A is not correct because multiple allelism is when three or more alleles for one gene exist. It does not mean all these alleles express themselves at once, just that they are possible alleles that can be inherited.

    C is not correct because linkage is when inherited genes on a chromosome stick together and produce certain trait combinations from this condition.

    D is not correct because nondisjunction is a genetic mutation caused by homologous chromosomes failing to separate during anaphase I of meiosis.

16. **D** is correct because large chromosomes, production of many offspring, and a short life cycle all are excellent traits for an organism to be the focus of a genetic study, such as the fruit fly.

17. **B** is correct because any individual that does not have the phenotype of the trait but has the genetics for the trait is considered a "carrier."

    A is not correct because a color-blind female would have the phenotype for color-blindness.

    C is not correct because if the female had the gene for color blindness on both X-chromosomes, she would be color-blind.

    D is not correct because of the same reason given in **C**.

18. **B** is the correct answer because 50% of the offspring will have blood type B as shown in the following Punnett square.

	B	B
A	AB	AB
o	Bo	Bo

19. **B** is not correct answer because the nitrogen base uracil is not found in DNA molecules only RNA molecules.

    A, **C**, and **D** are all components of DNA molecules.

20. **A** is correct because a codon is a group of three nitrogen bases in sequence.

    B is not correct because ribosomes are the site of protein synthesis in the cell.

    C is not correct because a nucleotide is the complete subunit of DNA and RNA molecules.

    D is not correct because peptides are the subunits of protein molecules.

21. **A** is the correct answer because translation is the name of the chemical process when the DNA molecule is copied by an mRNA molecule.

    B is not correct because transcription is when the mRNA molecule, tRNA molecule, and the ribosome come together to chemically create proteins.

    C is not correct because dehydration synthesis is when the union of smaller, simpler molecules with the removal of water creates complex molecules.

    D is not correct because hydrolysis is when complex molecules are broken up into smaller, simpler molecules with the addition of water.

22. **C** is the correct answer because three nitrogen bases (codon) are needed to code for each amino acid.

Short Response

23. DNA replication is when genetic information is duplicated in the nucleus of the cell for the chemical event of cell division. RNA transcription is when protein molecules are created in the cell from the codes brought to them by mRNA molecules and assembled in the ribosomes by tRNA molecules.

24. Looking at the following Punnett square, it can be viewed that a male offspring from this mating will have a 50% chance of being color-blind.

	X	X-
X-	XX-	X-X-
Y	XY	X-Y

25. A woman who has experienced genetic mutations of body cells would produce a child with no mutations. The information in the body cells is not inherited by the offspring. If the same woman had genetic mutations in the sex cells that were involved in the creation of a child, this individual would inherit these mutations from the mother.
26. In humans, more than two genes control the color of the skin. This is a condition called multifactorial inheritance. The wide range of skin colors in humans is the result of these many different genes being expressed in different combinations together.

    The pea plants studied by Mendel had only two genes that controlled the color of the seed pods. This is not multifactorial inheritance and the results are more defined and easy to predict.

Evolution

Today, the planet contains millions of forms of life. Scientists believe that the living organisms today represent less than 1% of all the life that has lived on this planet. Where have the rest of them gone? Why are they no longer here on this planet? Evolution is a branch of biology that attempts to look at how some forms of life have gone extinct, some forms of life have changed, and how some forms of life continue to thrive and have remained virtually unchanged for millions of years.

Lesson 5-1: Studying the Past

Organic evolution is the idea that living organisms that exist today have changed over time from earlier life that once lived on this planet. When many people think about life that once existed on this earth, the picture of huge dinosaur "skeletons" sitting arranged in some museum come to mind. The structures of these dinosaurs that can be viewed in most natural history museums are, in fact, remains of life that once roamed this earth. But these remains are but one type of fossil that can be studied when looking into the organic past. A **fossil** is the remains, or trace, of an organism that lived in Earth's past. ***Paleontologists*** are scientists who study fossils found in the earth. There are many different types of fossils that can be studied to learn about the organic history of Earth. Table 5.1 on page 144 shows some of the most common types of fossils that are studied.

Various Types of Fossils

Type of Fossil	Description
Hard parts of organisms	Bones, teeth, and shells are not as easily decomposed as many other parts of a living organism and can remain intact in certain environmental conditions for many years.
Petrification	The bones of animals and the cellulose within plants and trees can sometimes be replaced with minerals that preserve the shape of these structures.
Imprints	Tracks, tunnels, footprints, and leaf prints can be left in soft mud by ancient organisms that eventually harden into rock.
Amber	Small organisms, such as insects, that once roamed the earth can sometimes be trapped in the sticky resin of evergreen trees and sealed off from the decomposing agents of the environment.
Casts and molds	Sometimes rock forms around the structure of an organism, and when the organism decays, a mold of the organism remains in the rock.
Freezing	The preserved bodies of some prehistoric organisms have been uncovered in very cold parts of the world.

Table 5.1

Only a small percentage of all the organisms that have ever existed on this planet left fossil remains in the rock. Most of these fossils that do remain are found in **sedimentary rock**. Sedimentary rock is formed when an organism dies and becomes buried in mud. The fact that the fossil becomes buried quickly is important to prevent scavengers and decomposers from destroying the remains. More and more layers of sediment will eventually cover the buried organism. The heat and pressure generated from these many layers of sediment causes the sediment in the lowest layers to harden and preserve the organisms within.

How Old Is It?

In order to accurately study the past and how organisms have developed over time, scientists must be able to determine the age of the fossils they uncover in the earth. The two common ways in which scientists can determine the age of fossils are relative dating and radioactive dating.

As mentioned, most fossils are found in sedimentary rock that have formed in layers, or ***strata***, within the earth. **Relative dating** is when the fossils found deepest under the surface or at the lowest layer of the strata are considered the oldest, with the layers closest to the surface considered to be the most recent fossil remains. It is similar to finding a stack of papers or books on a desk, and determining that the book or paper at the bottom of the pile was laid down first. Relative dating allows paleontologists to compare the ages of two fossils found at different layers. Relative dating does not provide information about the exact age of either fossil.

Relative Dating

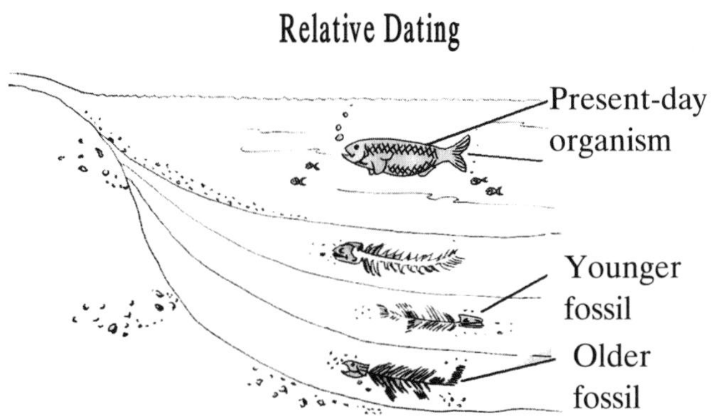

But by using remains or traces of past organisms called **index fossils**, it is possible to match the relative ages of sedimentary rocks in different parts of the world. One example of a good index fossil is an ancient extinct organism called the ***trilobite***. Some species of trilobites lived about 500 million years ago. Paleontologists can use this information to date other fossils they may find in places around the world. Any sedimentary rock found throughout the world that contains this species of trilobite will be around 500 million years old. This means that fossils found in this immediate layer with the trilobites or just below or above that layer can be relatively dated.

Chemicals found in the rocks and remains of life left on the planet contain radioactive isotopes. **Radioactive isotopes** are elements whose nuclei break apart or decay and emit energy in the form of radiation. Many times when these elements break apart, they actually become new elements. This tells you that these elements must be losing protons. The rate at which these radioactive elements break apart is called **half-life**. Half-life is the amount of time it takes for half of the elements in a newly created sample of rock or fossil to break apart into new elements. Any type of chemical or physical force in nature does not affect half-life. This means that if you know how long the half-life of a particular radioactive element is and you know how much of the element has already decayed in the sample, you can accurately date when the rock or fossil first formed. There is a radioactive isotope of carbon called ***carbon-14***. When carbon-14 is taken into plants or animals during the process of nutrition, this element becomes part of their structure. When these organisms die, they will no longer take in carbon-14. The half-life of carbon-14 is 5,730 years. This means that if you have a sample of carbon-14, in 5,730 years you will have only half that original amount. If a scientist knows the amount of carbon-14 thought to be present in the organism at the time of death, and then compares this to the amount of carbon-14 in the fossil remains he uncovers, the age of the fossil can be determined.

Example

A scientist finds a petrified tree fossil that contains 20 grams of carbon-14. By understanding the possible chemical make-up of the tree when it was alive, the scientist determines that at the time of death the tree had 80 grams of carbon-14 in its cells. The scientist knows that the original amount of 80 grams of carbon-14 at the time of death will decay to half that amount in 5,730 years. The amount of carbon-14 in the fossilized tree is now 20 grams, which tells the scientist that the tree fossil has gone through two half-lives.

Solution:

The amount of time for one half-life (5,730 years) is multiplied to the number of half-lives recorded (2), which means the tree died about 11,460 years ago.

••••••

Table 5.2 shows some radioactive isotopes and their half-lives. Depending on the age of something that needs to be dated accurately, various isotopes, because of their half-life, would be more or less helpful.

Radioactive Isotope Half-Lives

Isotope	Half-Life	Isotope	Half-Life
Uranium-238	4.5 billion years	Plutonium-239	24,300 years
Potassiu-40	1.26 billion years	Carbon-14	5,700 years

Table 5.2

History of Life on Earth

During the study of the Earth, distinct rock layers containing a variety of very different fossils were discovered, this led scientists to divide the history of life on Earth into large units called ***eras***. Table 5.3 shows the four major eras of life on the planet.

The Time Line on Earth

Era	Major Event	Era Started This Many Years Ago
Precambrian	Life on the planet was confined to bacteria, simple algae, and some invertebrates.	3.8 billion
Paleozoic	Age of Invertebrates Age of Fishes Age of Amphibians	570 million
Mesozoic	Age of Reptiles (Dinosaurs) Rise of Flowering Plants	245 million
Cenozoic	Age of Mammals Rise of Primates and *Homo sapiens*	65 million

Table 5.3

How did life start?

The cell theory presented in Chapter 3 includes the concept: "all cells come only from other cells." In essence, this means that life comes from life. An obvious exception to this concept would have to be the first cell on this planet or the first living organism. There are some hypotheses addressing the beginning of life on this planet. Many of these ideas do not have a great deal of data to support them, but may be as close as we get to a possible answer to this very intriguing question.

In the 1930s a Russian scientist, Alexander Oparin, hypothesized that life began in the oceans of this planet around 3.6 billion years ago. Oparin suggested that small organic molecules were formed from the sun's energy, lightning, and substances found in the early atmosphere. He then stated that rain probably washed all these materials into the ocean from the atmosphere and they formed simple organic molecules.

In 1953, two American scientists, Stanley Miller and Harold Urey, attempted to test Oparin's hypothesis by recreating the conditions of ancient Earth. Miller and Urey used chemicals such as water, hydrogen, ammonia, and methane because they were believed to be present in the ancient atmosphere. They continually vaporized and then condensed these chemicals and intermittently sent electric sparks into this mixture. After only about one week a few different kinds of amino acids were created. Amino acids are the building blocks of proteins, which are the structural components of living things. Today it is believed that the chemicals used by Miller and Urey were probably not the ones present in Earth's early atmosphere; but similar experiments with new chemicals also produced organic compounds.

Heterotrophic Hypothesis

At first, many scientists hypothesized that the first living organisms on this planet must have been autotrophs. This would seem like a reasonable idea because plants are the backbone of present-day ecosystems. But the chemical process of photosynthesis is a very complex reaction and also requires oxygen, an element not in abundance in early Earth's atmosphere.

The newest and most widely supported hypothesis regarding the first forms of life on this planet suggests that they were anaerobic heterotrophs. An anaerobic heterotroph would be an organism that ingested already present organic molecules from an oxygen-starved environment. These anaerobic heterotrophs and their chemical reactions of nutrition would have, through time, released enough carbon dioxide into the atmosphere to provide an environment that could have supported the anaerobic autotrophs. These anaerobic autotrophs and their chemical processes of nutrition would have, over time, released enough oxygen into the atmosphere to provide an environment suitable for aerobic autotrophs and heterotrophs.

The main ideas of the heterotrophic hypothesis were formulated over a 20-year period by the work of J.B.S. Haldane, A.I. Oparin, and Harold Urey.

Spontaneous Generation

For thousands of years, many forms of life were thought to be the result of ***abiogenesis***, also known as spontaneous generation. Spontaneous generation's doctrine is basically that organic life could arise from inorganic matter. Across most of the world there were recipes for creating life.

Example

One example of a recipe for life is:

Recipe for creating mice:

1. Get sweaty rags.
2. Wrap them around husks of wheat.
3. Leave this in an open jar for three weeks.
4. Mice will be created from the mix.

••••••

In the late 1600s, an Italian scientist by the name of Francesco Redi set out to disprove the notion of spontaneous generation. During the time of Redi, many people believed that maggots arose from rotting meat. Redi believed otherwise and designed the following experiment to prove his contrary idea for where the maggots arose.

Francesco Redi's experiment showed that it was not the rotting meat that was causing the production of flies, but that flies were laying eggs on the meat or the net (too small to be seen), and these eggs were then hatching into new flies.

The experiment designed by Redi was also one of the first scientific experiments recorded that featured a ***control group***. The closed jar containing the rotting meat was the control group used as a basis of comparison in the experiment.

Francesco Redi Experiment

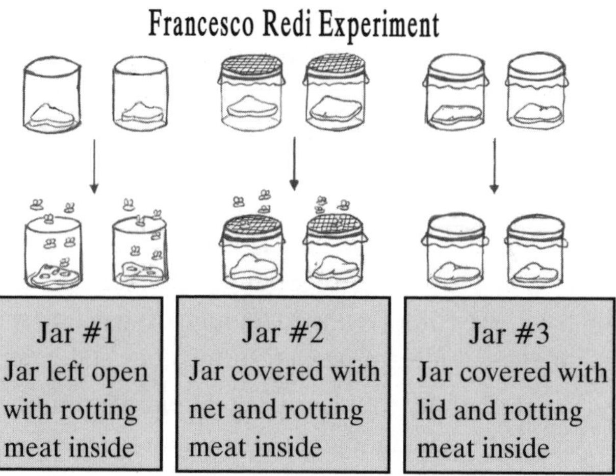

Practice Section 5-1

1. Which of the following would not be considered a fossil?
   a) a bird's footprint
   b) an insect in amber
   c) an axe blade
   d) the tusk of an ancient elephant

2. Most fossils are found in
   a) igneous rock
   b) metamorphic rock
   c) sedimentary rock
   d) tar pits

3. Relative dating is when fossils are dated according to
   a) their placement in the strata of the planet
   b) the amount of undecayed radioactive atoms inside their structure
   c) how large the preserved fossil is in mass
   d) what percentage of carbon-14 is found inside the structure

4. What era featured the rise of *Homo sapiens*?
   a) Cenozoic b) Paleozoic c) Precambrian d) Mesozoic

5. The theory of the spontaneous generation of life from inorganic materials was seriously threatened by the experimental work of which scientists?
   a) Charles Darwin
   b) Stanley Miller
   c) Francesco Redi
   d) Robert Hooke

6. Petrifaction is best explained as the process by which
   a) organic materials are converted into humus
   b) minerals are dissolved and carried to the surface
   c) the tissue of the organisms is gradually replaced with minerals
   d) the tissue of the organisms is frozen in ice

7. When did life first appear on this planet?

8. What is "spontaneous generation"?

9. How did the work of Stanley Miller add to our knowledge base for the origin of life on this planet?

Lesson 5-2: Scientific Evidences of Evolution

Scientists study the fossil records and living organisms existing on the earth today to show the changes that have occurred to living organisms over the past 3.5 billion years. We will now look at some of the ways scientists have sought to prove the process of evolution.

Evidence From Fossils

The fossil uncovered by geologists and paleontologists show that life on Earth has evolved over the course of many millions of years. Relative dating of fossils in the deepest strata shows that organisms in these layers were simpler in structure than fossils in strata layers closer to the surface. There are many organisms that the fossil records show very clearly how their forms developed over time.

Evidence From the Study of Cells

All living organisms are made up of the same structures: cells. Many cells that exist on this planet are very similar, containing many of the same organelles and conducting the same chemical processes. These facts show an underlying unity among all living organisms, suggesting a common ancestor. The study of cell structures and evolution is called ***cytology***.

Evidence From DNA

The structure of DNA and how it functions in all living things is remarkably similar. Many of the amino acid sequences that create proteins are very similar in organisms that show a close relationship, while organisms that are not as close show more variation in how these proteins are formed.

Evidence From Anatomy

Scientists can dissect and study the similarities between various organisms. How similar the structures of two organisms are can be used to show a common ancestor. When scientists compare similar structures, they have to keep two important factors in mind called homologous and analogous. **Homologous structures** are structures that are being compared that have the same evolution and origins, but not necessarily the same function. For example, the human arm and a whale's flipper are structures that have evolved from a common ancestral structure, but now these structures are used to do different tasks for the respective organisms.

Analogous structures are structures present in two different organisms that have the same function but are structurally different and originate from different origins. For example, the wing of a mosquito and the wing of a bird are both used for flying, but these structures are made up of different materials and originate from different evolutionary paths.

Homologous Structures

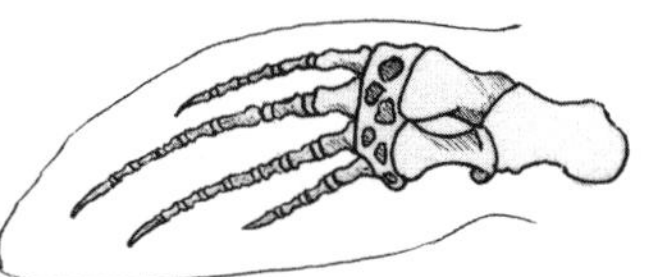

Whale flipper

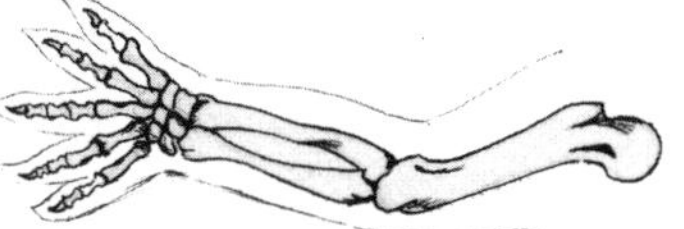

Human arm

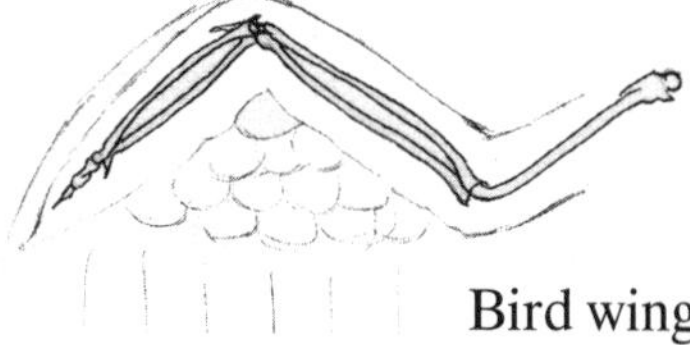

Bird wing

Evidence From Embryology

Comparison of the embryos of some vertebrates shows great similarities. For example, in organisms such as chickens, pigs, and humans, the developing embryos all have gill slits, segmented backbones, C-shaped bodies, and tails. This shows that these organisms must have had a common ancestor.

Evidence From Vestiges

A ***vestige*** is a structure found in an organism that does not have any use, but may have had a function in the ancestors of the organism. One very well-known example of a vestige structure is the human appendix. The appendix is a small pouch-like structure found at the connection point between the small and large intestines. In some herbivores, this pouch is larger and functions as an additional area for food digestion. This shows our connection to many of these herbivore animals.

Practice Section 5-2

1. Which of the following is not an example of physical evidence used to support the theory of evolution?
   a) Fossil evidence of ancient life forms
   b) Comparing cell structures of organisms
   c) Comparing anatomies of organisms
   d) All are used
2. What is a vestige?
   a) The wing of any ancient bird
   b) A body structure that is no longer used
   c) A method of dating ancient fossils
   d) The leg structure of the first land animal
3. Two structures that have the same origins and evolution but have different present-day functions are called
   a) homologous structures
   b) analogous structures
   c) vestiges
   d) cytological structures
4. Which of the following are examples of analogous structures?
   a) The human arm and the flipper of a whale.
   b) The human arm and the wing of a bird.
   c) The wing of a bird and the wing of a bee.
   d) All of these are analogous structures.

5. Many amino acid sequences that create proteins are similar in some organisms. This is what type of evolutionary evidence?
   a) fossil b) DNA c) vestiges d) embryology
6. Cytology is the study of
   a) ancient fossils b) cells c) vestiges d) varying anatomies
7. What is the scientific study of cells called? How can this process aid in the study of evolution?
8. What is the difference between homologous and analogous structures?
9. What is a vestigial structure? Give an example of one.

Lesson 5-3: Theories of Evolution

LaMarck: Use It or Lose It!

In the early 1800s, a French biologist, **Jean Baptiste Lamarck**, introduced his ideas about the mechanisms involved in evolution. Lamarck felt that organisms change because they need to respond to stimuli from their environments. His theory about evolution states that if a living thing uses a particular structure more, it will develop more; and if a living thing uses a structure less, it will become weaker, less developed, and will eventually disappear. Lamarck also believed that once these structures change, based on use or disuse, that these characteristics could then be passed on to their offspring. For example, the ancestor of the giraffe was believed to have a short neck, but from years of many organisms stretching to reach the high leaves on trees, their necks became longer and longer.

A scientist by the name of **August Weismann** tested Lamarck's theory of evolution by conducting evolutionary tests on some organisms. Weismann tested Lamarck's theory on mice in his labs. He had their tails cut off and then they were mated with other mice that had their tails cut off. This was done for many generations, but no mice were ever born without tails. Lamarck's ideas about evolution could not be proven true by experimental procedures.

Darwin: Survival of the "Fit"

Charles Darwin (1809–1882) is considered by many scientists to be the "father" of evolution. Charles Darwin produced a large collection of papers during his life, with his most famous work, *On the Origin of Species*, being published in 1859.

Following his graduation from Cambridge, Darwin took an extended sea voyage exploration. He became the ship's naturalist, where he studied and took notes of the many animals and plants he discovered in his journey along the coastline of South America.

When Darwin returned to England from this expedition, he analyzed all the notes he had taken and began reading about other research that had been done by fellow scientists. All these ideas that he accumulated, based on the notes from his journey and the research he had done in England, became the basis for his theory of evolution, which appeared in *On the Origin of Species*.

The ideas of Darwin's theory can be summarized as follows:

- Most species of life on Earth normally produce larger numbers of offspring than can possibly survive in the environment.
- A particular environment cannot usually support all the organisms born into that environment, because of limitations on food and shelter.
- All organisms born of a particular species differ slightly in size, color, type of claws, and so on. Some of these differences found in organisms provide them with an advantage in coping with their environment.
- The organism that is best suited for its environment will survive and produce the most offspring. The organism that is least fit for its environment will not survive as long and will produce few to no offspring.
- A new species can develop when advantageous differences accumulate in a current species. This occurs when there are so many structural and behavioral differences that the organisms will have changed into an organism that cannot mate with the original form of the organism.

••••••

The explanation of how these variations or differences within a species could arise was one limitation to Darwin's theory of evolution that was discussed in his book *On the Origin of Species*. During Darwin's time, the science community knew nothing of the gene or how it affected the traits of an organism.

A scientist by the name of **Hugo Vries** (1848–1935) was able to explain at least why some of these variations occurred. In the early 1900s, Vries stated that new species developed from sudden changes in their genes, called ***mutations***. Mutations can be helpful or harmful to the individual organism. Organisms that accumulate helpful mutations will eventually develop into new species. For example, an elephant born with very large tusks, which proves to be beneficial for both mating and food gathering, will likely survive and produce more offspring than an elephant with smaller tusks. This is why many African elephants have such large tusks. Recently, though, because of intensive hunting of African elephants for their long ivory tusks, the once beneficial trait of having long tusks is becoming harmful. Poachers target elephants with the longest tusks in a herd because this action will bring potentially bring the highest payoff from the sale of the ivory tusk. In many populations of these animals, elephants with smaller tusks are now more common than elephants with longer tusks. The higher frequency of poaching elephants with longer tusks has resulted in the decrease of animals with this particular attribute in the population.

Practice Section 5-3

1. Which idea is not part of Darwin's idea of evolution?
   a) The environment will select the most "fit" organism to survive.
   b) Most organisms produce more offspring than can possibly survive.
   c) Mutations can cause living organisms to change.
   d) Variations in an organism will be inherited.
2. The theory that characteristics acquired during the life of a parent organism will be passed down to its offspring was proposed by
   a) Darwin b) Weinberg c) Lamarck d) Linneaus
3. What is the name of the famous book written by Charles Darwin that explains his ideas about evolution?
   a) *Changing Animals* c) *Darwin's Great Proposa*
   b) *On the Origin of Species* d) *Evolution of Life*
4. If a man were to obtain huge muscles from lifting weights in a gym on a regular basis, which scientist would say that the offspring of this man would be born with large muscles?
   a) Darwin b) Gould c) Lamarck d) Miller
5. According to Charles Darwin, organisms that will be the most successful biologically are organisms that
   a) are best adapted to their environment
   b) are the largest in the population
   c) reproduce the slowest
   d) feed on other organism in the ecosystem
6. Many changes that occur in living organism are the results of mutations. This idea was first put forth by
   a) Charles Darwin c) Hugo Vries
   b) Jean Baptiste Lamarck d) Jay Gould
7. What is meant by the evolutionary idea of "Use it or Lose it"?
8. What is a mutation? How is this process important in evolution?
9. What is meant by the idea "survival of the fittest" proposed by Charles Darwin?

Lesson 5-4: Human Evolution

Modern humans, or ***Homo sapiens***, arose on this planet about 50,000 years ago. Like all other animals, humans have changed over time on this planet. Each year, paleontologists learn more about the evolution and development of modern man through their new discoveries of fossils in the field. Unfortunately, at this time, very few fossils of our ancestors have been discovered, but enough have been unearthed to begin to form the evolutionary history of man.

The term ***hominid*** is used to represent members of the taxon found under primates in the biological classification system. Hominids include modern man (*Homo sapien*), as well as many extinct manlike organisms represented by fossil records.

One of the earliest known hominids that stood on two feet in the human family tree was ***Australopithecus afarensis***. In 1974, this species became known with the discovery of a collection of fossils by a team led by the paleontologist Donald Johanson. The famous Beatles song "Lucy in the Sky with Diamonds" was playing at camp the day of the discovery so the scientists named the discovered organism "Lucy." Fossils from about 13 members of this species were collected in the immediate area of the "Lucy" find. This led paleontologists to believe that *A. afarensis* lived in small social groups. *A. afarensis* had brains about the size of modern-day chimpanzees and stood about 3 feet tall.

The earliest organism that could be classified as human is ***Homo habilis***. This species of humans arose on Earth about 2 million years ago. This was the first human species believed to use stone tools. *Homo habilis* was the first of several species in our genus to arise around this time period.

Homo ergaster appeared on the planet about 1.5 million years ago. *Homo ergaster* is believed to be the first members of our species to venture out of Africa and invade parts of Asia and Europe. *Homo ergaster* is sometimes referred to as the "naked ape." Three species of humans are believed to have developed out of this species, including: ***Homo erectus***, ***Homo sapien,*** and ***Homo neanderthalensis***. Only the species *H. sapien,* developing through a species called ***Homo heidelbergensis,*** would remain about 1 million years later.

The fossil record shows that the species of modern man, *Homo neanderthalensis,* probably arose on this planet about 80,000 years ago. This species of modern man is the closest extinct member of the human lineage, and it went extinct for unknown reasons about 40,000 years ago. Arising at around the same time as *Homo sapiens neanderthalensis*, was a second subspecies of modern humans named Cro-Magnon people. Cro-Magnon people are the prehistoric representatives of the species *Homo sapiens sapiens*, which means this group is considered to be the same species as modern man. There are some scientists who now believe that *Homo sapiens* may have been responsible for the extinction of the second subspecies *neanderthalensis*.

What has become increasing clear to paleontologist, as they have uncovered more and more of *Homo sapiens*' ancestors, is that there was not a direct line of evolution. If we start with the earliest walking hominid "Lucy" and move to the present, many offshoot species related to humans were tried and failed over this great time period. For example, there lived a group of hominids now placed in a separate genus called ***Paranthropus***. These organisms were larger in size than modern humans and had thick grinding teeth used to eat tough plant material such as ground tubers. Two members of this dead-end genus include ***P. robustus*** and ***P. boisei***. Table 5.4 shows a partial family tree based on fossil evidence of the human ancestry. The species shown are those believed to be in the direct line to modern man.

The Human Family Tree

Genus and Species	Time on This Planet
A. afarensis	3.5–3.0 million years ago
A. garhi	3.0–2.0 million years ago
H. habilis	2.0–1.5 million years ago
H. ergaster	1.5–1.0 million years ago
H. heidelbergensis	1.0–50,000 years ago
H. sapien	50,000–present

Table 5.4

The general evolutionary trend in body structure from *A. afarensis* to *H. sapien* would be:

1. An increase in body size.
2. An increase in brain size.
3. Diminished hair on the body.
4. An increased ability to modify the environment.

••••••

Practice Section 5-4

1. Which of the following is not a characteristic of the genus *Homo*?
   a) bipedal movement
   b) tool use
   c) large brain size
   d) large sharp canines

2. Which of the following members of the human ancestry arose first on this planet?
   a) *Homo sapien* b) *Homo erectus* c) *A. afarensis* d) *Homo ergaster*

3. Which of the following statements is true about human evolution?
   a) The first hominids arose in North America.
   b) The brain case of hominids has gradually decreased in size.
   c) The only hominid to use tools is *Homo sapien.*
   d) The first hominids arose in Africa.

4. The hominid named "Lucy" was a member of which group?
   a) *H. sapien* b) *H. ergaster* c) *A. afarensis* d) *H. habilis*

5. Which of the following organisms are not considered by paleontologists to be part of the direct line to modern man?
   a) *H. habilis*
   b) *A. afarensis*
   c) *H. heidelbergensis*
   d) *P. boisei*

6. Modern man (*Homo sapien*) arose on this planet
   a) 1 million years ago
   b) 3 million years ago
   c) 200,000 years ago
   d) 50,000 years ago

Terms From Chapter 5		
analogous structures	hominid	organic evolution
cytology	homologous structures	radioactive dating
era	index fossil	radioactive isotope
fossil	mutation	relative dating
half-life		vestige

Chapter 5: Exam

Matching Column for Types of Fossils

Match the type of fossil with the best description of that particular type of fossil. You may use an answer choice more than once or not at all.

Type of Fossil

1. petrification
2. imprints
3. amber
4. freezing
5. hard parts
6. casts and molds

Description

a. When rocks form around the structure of an organism leaving an empty space inside the rock in the shape of the decayed organism.

b. The frozen preserved body parts of an organism.

c. Bones, teeth, and other not easily decomposed parts of the organism.

d. Small organisms can become trapped and preserved in this sticky resin from evergreen trees.

e. When the bones of animals and the cellulose of plants are replaced with minerals.

f. Tracks, tunnels, footprints, and leaf-prints left in the soft mud and turn into hard rock.

Matching Column for Types of Fossils

Match the physical evidence of evolution with the best description of that particular type evidence. You may use an answer choice more than once or not at all.

Physical Evidence

7. Anatomy
8. vestiges
9. fossils
10. embryology
11. cytology
12. DNA evidence

Description

a. A structure found inside an organism that does not have any used now, but did in the ancestors of that organism.

b. Comparing the similarities and differences between amino acid sequences in two organisms.

c. Dissecting and studying the similarities and differences between organs and body structures of two organisms.

d. The remains of deceased organisms that are studied.

e. Comparisons of the early developmental stages of an organism.

f. Comparing and contrasting cell structures found within an organism.

Multiple Choice

13. The study of how organisms have changed structurally and biochemically from earlier forms of life.

    a) ecology b) evolution c) embryology d) paleontology

14. Relative dating is used to determine

    a) the age of a fossil based on radioactive isotope percentages

    b) the age of a fossil based on where in the strata it is located

    c) the age of a fossil based on its structural composition

    d) the age of a fossil based on its chemical makeup

15. If the half-life of a particular element is five years, this means that in five years

    a) the element will be totally gone

    b) the element will have doubled in mass

    c) half of the element will have changed

    d) none of the above

16. Which of the following is a basic idea of the heterotrophic hypothesis?
    a) More complex organisms led to simpler organisms.
    b) Multicellular autotrophic organisms led to one-celled heterotrophic organisms.
    c) No life appeared on this planet until there was sufficient oxygen in the atmosphere.
    d) Autotrophic chemical processes added oxygen to the atmosphere.
17. Which of the following statements is in accordance with Darwin's theory of evolution?
    a) Organisms produce more offspring than can usually survive.
    b) A particular environment cannot support all the organisms born into that environment.
    c) The organism best suited for that environment will have the best chance at survival.
    d) All of the answers.
18. Which of the following scientists believed that characteristics acquired by the parent organism, during life, could then be passed on to their offspring?
    a) Charles Darwin
    b) Jean-Baptiste Lamarck
    c) August Weismann
    d) Hugo Vries
19. Which of the following members of the human ancestry arose on this planet first?
    a) *Homo sapien*
    b) *Homo ergaster*
    c) *Australopithecus afarensis*
    d) *Homo neanderthalensis*

Short Response

20. A sample from a preserved tree that was uncovered today contains 24 grams of the isotope carbon-14. Scientists working on this specimen have calculated that at the time of death, this tree contained 96 grams of carbon-14. If the half-life of carbon-14 is 5,730 years, how old is this preserved tree?
21. Explain what Charles Darwin meant when he stated that certain organisms are more "fit" than other organisms in a particular environment. How can this eventually lead to speciation?
22. Explain how the following statement explains organic evolution."Individual organisms do not adapt to the environment but populations can."

Answer Key

Answers Explained Section 5-1

1. **C** is the correct answer because an axe blade would not be considered a fossil, because an axe blade is a rock not the remains of a living organism.

   A, **B**, and **D** are not the correct answer because a bird footprint, an insect trapped in amber, and the tusk of an ancient elephant would be considered fossils, because they are all either actual remains of living organisms or the structural mark these living organisms have left in their environment.

2. **C** is the correct answer because most fossils are found in sedimentary rock. Sedimentary rock forms from pieces of organic debris and small rocks that are washed into areas that contain water and then slowly harden. This is a good process for the preservation of organic remains.

   A is not correct because igneous rock is formed from molten rock from deep in the Earth that cools at the surface. This is not a good process for preserving the remains of life.

   B is not correct because metamorphic rock is formed deep under the earth's surface where heat and pressure harden soft materials. This is not a good process of preserving the remains of life.

   D is not correct because tar pits do have the ability to preserve fossils, but larger numbers of ancient remains have been found in sedimentary rock than tar pits.

3. **A** is the correct answer because relative dating is when fossils are dated according to the level below the surface of the Earth they are found inside.

   B is not correct because radioactive dating is accomplished by measuring the amount of undecayed atoms inside the fossil.

   C is not correct because the size of the fossil does not usually say anything about its age.

   D is not correct because the percentage of carbon-14 found inside the fossil is a form of radioactive dating.

4. **A** is the correct answer because the era on Earth that witnessed the rise of *Homo sapien* is the Cenozoic.

   B is not correct because Precambrian era was dominated by single-celled life forms.

   C is not correct because Paleozoic era was dominated by fish, amphibians, and invertebrates.

   D is not correct because Mesozoic era was dominated by reptiles (The Age of Dinosaurs).

5. **C** is the correct answer because the work of Francesco Redi damaged the idea of spontaneous generation. Redi's work showed that many organisms that looked like they were created from inorganic beginnings actually were not.

   A is not correct because Charles Darwin is noted for his work on the theory of evolution.

   B is not correct because Stanley Miller is noted for his work on how life first arose on this planet.

   D is not correct because Robert Hooke is noted for his work with the compound microscope and cell studies.

6. **C** is the correct answer because the tissue of a dead organism is gradually replaced with minerals in the process of petrifaction.

   A is not correct because organic material converted into humus is the chemical process of decomposition.

   B is not correct because minerals that are dissolved and carried to the surface would be a form of run-off.

   D is not correct because freezing is when the tissue of a dead organism is frozen in ice and preserved.

7. It is estimated that life began on this planet about 3.6 to 3.8 billion years ago.

8. The idea of spontaneous generation is an obsolete theory regarding the origin of life from inanimate matter. For example it was believed according to this theory that organisms like mice could arise from piles of old rags. The scientific work of Francesco Redi did much to end the belief of life arising from inorganic matter.

9. In 1953, Stanley Miller designed and conducted an experiment to simulate the conditions of the atmosphere before life arose on this planet. Miller combined methane, ammonia, hydrogen gas, and water vapor and applied an electrical current to this chemical soup for a series of days. The results of this experiment proved that certain organic chemical compounds could be created from these inorganic chemicals under the right conditions.

Answers Explained Section 5-2

1. **D** is the correct answer because **A**, **B**, and **C** are correct. Fossil remains, comparison of cell structures, and comparison of anatomies, are examples of physical evidence supporting the theory of evolution.

2. **B** is the correct answer because a vestige is a body structure that is no longer used in the present form of the organism but had a use in the ancestral form.

A is not correct because the wings of ancient birds and the wings of a modern bird both had useful functions for the organisms in question.

C is not correct because the method of dating ancient fossils is either relative or radioactive dating.

D is not correct because the leg structure of the first land animal is a structure that would gain increased use and importance as the organism evolved.

3. **A** is the correct answer because structures that are found in different organisms that had the same function in the ancestral form but have different present day functions are called homologous structures. An example of this is the arm of a bird and the arm of a human.

   B is not correct because vestiges are body structures that are no longer used in the present form of the organism but had a function in the ancestral form of the organism.

   C is not correct because analogous structures are structures found in two different organisms that have the same present-day functions but had different ancestral origins.

   D is not correct because cytological structures are structures found in the make-up of cells.

4. **C** is the correct answer because the wing of a bird and wing of a bee are analogous structures. The wings of the bird and the bee have the same present day function, which is flight, but did not originate from the same structures in the body in the ancestral organisms.

   A is not correct because the human arm and the flipper of the whale are considered to be homologous structures. Homologous structures have different present day functions but the same evolution from an ancestral form.

   B is not correct because the human arm and the wing of a bird are considered to be homologous structures.

5. **B** is the correct answer because when amino acid sequences and protein formation is being compared in organisms, the evolutionary evidence is the DNA of the organisms.

   A is not correct because a fossil is the remains or trace of an organism that lived in Earth's past.

   C is not correct because vestiges are body structures that are no longer used in the present form of the organism but had a function in the ancestral form of the organism.

   D is not correct because embryology is the study and comparison of early development between two organisms.

6. **B** is the correct answer because cytology is the study of cells.

   A is not correct because ancient fossils are the remains or traces of organisms that lived in Earth's past.

   C is not correct because vestiges are body structures that are no longer used in the present form of the organism but had a function in the ancestral form of the organism.

   D is not correct because anatomies are the entire body tissue structures of various organisms.

7. Cytology is the study of the structure and evolution of cells. Since all living organisms are composed of cells made of roughly the same structures, this underlying unity among all living organisms suggest a common ancestor.

8. Homologous structures are structures being compared on two different organisms that have the same evolution and origins, but not always the same function. An example of homologous structures would be a dog's paw and a human hand. The bones in a dog's paw and the bones in a human hand have a common evolutionary history, but are now used for two different things. Analogous structures are structures being compared on two different organisms that have the same present function but have different evolutionary paths. An example of analogous structures would be the bones in the wings of a bat and the bones in the wing of a bird. The bones in each of these organisms are now used for flight, but their evolutionary history is not the same.

9. A vestigial structure is a characteristic of an organism that seems to have lost all or most of its original function through evolution of non-use. Baleen whales have vestigial legs bones, unused, buried deep within the back of the organism.

Answers Explained Section 5-3

1. **C** is the correct answer because the idea that is not part of Darwin's idea of evolution is mutations can cause living organisms to change. The idea of mutations as the agents of change causing evolution was first proposed by Hugo Vries not Charles Darwin.

   A, **B**, and **D** are not correct because the ideas that the environment will select the most fit organism, that most organisms produce more offspring than normally can survive, and variations present in organisms will be inherited are all ideas put forth in Darwin's book *On the Origin of Species*.

2. **C** is the correct answer because Lamarck proposed the idea that traits acquired during life by the parent organism could be inherited by the offspring. This idea was later proven to be false.

A is not correct because Darwin's theory for evolution was the most "fit" organism will survive and produce the most offspring for the next generation.

B is not correct because Weinberg worked on gene frequencies in ecological population groups.

D is not correct because Linneaus devised the binomial nomenclature classification system.

3. **B** is the correct answer because the book written by Charles Darwin that explains his ideas about evolution is titled *On the Origin of Species*.

   A, **C**, and **D** are not correct because *Changing Animals*, *Darwin's Great Proposal*, and *Evolution of Life* are all made up titles by the author of this review book.

4. **C** is the correct answer because Lamarck believed acquired traits, such as large muscles from lifting, could be inherited from the parent organism.

   A is not correct because Darwin believed the organism born with the best traits for the particular environment will survive and produce the most offspring for the next generation.

   B is not correct because Gould was a world renowned paleontologist and author.

   D is not correct because Miller experimented on the theory of how life began on this planet.

5. **A** is the correct answer because Charles Darwin believed that organisms that are best adapted to their environment will be the most biologically successful. Successful in biological terms means the ability to produce large numbers of offspring for future generations.

   B is not correct because the largest organisms are not always the organisms that are best suited to their environment and able to produce offspring in large numbers for future generations. For example, African elephants are endangered animals.

   C is not correct because organisms that reproduce very slowly are usually at a disadvantage because they have a difficult time getting large numbers of offspring into the future.

   D is not correct because autotrophic and heterotrophic organisms will be successful based on their ability to exist in their environment, not usually because of their general nutritional requirements.

6. **C** is the correct answer because the idea of mutations was first proposed by Hugo Vries.

   A is not correct because Charles Darwin's work centered on ideas concerning evolution. At Darwin's time, the idea of mutations was not known.

B is not correct because Jean Baptiste Lamarck proposed the idea of acquired traits from the parent being inherited by the offspring.

D is not correct because Jay Gould was a world renowned paleontologist and author.

7. "Use it or Lose it" is the idea that an organism can pass on characteristics that it acquired during its lifetime to its offspring. Also an individual organism will lose characteristics they do not require and pass this onto their offspring. For example if a blacksmith developed very strong arm muscles from his work, then his offspring, once fully grown, would acquire the same strong muscles in the arm.
8. Mutations are changes in a genomic sequence. Mutations are caused by radiation, viruses, and certain chemicals, as well as errors during the normal process of meiosis. Organisms that accumulate beneficial mutations can eventually develop into new species.
9. Survival of the Fittest means that the organism that is best suited for its environment will survive and produce the most offspring, while the organism that is less suited for its environment will not be able to produce as many offspring for the next generation.

Answers Explained Section 5-4

1. **D** is the correct answer because large, sharp canines is not a characteristic of the genus *Homo*.

   A, **B**, and **C** are not correct because bipedal movement, the use of tools, and a large brain size are all important characteristics that distinguish this genus from others.
2. **C** is the correct answer because the first member of the human ancestry from the group listed to arise on this planet was the species *A. afarensis*. This species arose on this planet about 3–3.5 million years ago.

   A is not correct because *Homo sapien* is the classification for present-day human beings and members of this present form are believed to have arose on this planet about 50,000 years ago.

   B is not correct because *Homo erectus* is an ancient form of hominid, not in the direct line of modern humans, and arose on this planet probably about 1.5 million years ago.

   D is not correct because *Homo ergaster* is an ancient form of modern man that arose on this planet about 1.5 million years ago.
3. **D** is the correct answer because the true statement about human evolution is that the human species is believed to have arisen in Africa.

A is not correct because the human species is believed to have arisen in Africa, not North America.

B is not correct because the brain case of hominids has gradually increased in size, not decreased.

C is not correct because there were ancient members of the hominid line before modern humans believed to have used tools such as *Homo habilis.*

4. **C** is the correct answer because the famous hominid dubbed "Lucy" by the group of paleontologist that discovered her remains is a member of *A. afarensis.*

   A is not correct because *H. sapien* is modern man.

   B is not correct because *H. ergaster* is an ancient form of modern man that lived about 1.5 million years after the "Lucy" species.

   D is not correct because *H. habilis* is an ancient form of modern man that lived about 1 million years after the "Lucy" species.

5. **D** is the correct answer because paleontologist now believe that the robust hominid remains of the species *P. bosei* are not in the direct line of human evolution.

   A, **B**, and **C** are not correct because *H. habilis*, *A. afarensis*, and *H. heidelbergensis* are all believed to be ancient forms in the direct line to modern man.

6. **D** is the correct answer because modern man is believed to have arisen on this planet about 50,000 years ago.

   A, **B**, and **C** are not correct because in the time period from 200,000 to 3 million years ago, there were members of the hominid line but modern man had not yet developed.

Answers Explained Chapter 5 Exam

Matching Column

1. E	4. B	7. C	10. E
2. F	5. C	8. A	11. F
3. D	6. A	9. D	12. B

Multiple Choice

13. **B** is the correct answer because evolution is the study of how organisms have changed from ancient forms to present-day forms.

    A is not correct because ecology is the study of how living organisms interact with other organisms and the environment.

C is not correct because embryology is the study of early development of an organism.

D is not correct because paleontology is the study of extinct organisms and their impressions left in the environment.

14. **B** is the correct answer because relative dating is when a fossil is dated according to where in the strata it is located and its relationship to other fossils located in the strata.

    A, **C**, and **D** are not correct because the amount of radioactive isotopes, the structural composition, and the chemical make-up of the fossil would all be forms of radioactive dating.

15. **C** is correct because if the half-life of an element is five years, this means that in five years, half of the atoms in this sample of the element will have changed to another form.

    A is not correct because according to the Law of Conservation of Mass, the atoms that make up an element can not just disappear.

    B is not correct because according to the Law of Conservation of Mass, the atoms that make up an element can not double in mass without the addition of mass to the original sample.

16. **D** is correct because the heterotrophic hypothesis has as one of its postulates that the rise of autotrophic organisms led to an increase in the amount of oxygen in the early atmosphere.

    A is not correct because simple organisms have led to more complex organisms.

    B is not correct because the heterotrophic hypothesis states that heterotrophic organisms led to autotrophic organisms.

    C is not correct because evidence suggests that life existed on this planet before large amounts of oxygen appeared in the atmosphere.

17. **D** is correct because according to Darwin's theory of evolution, all of the following are correct: organisms produce more offspring than can usually survive, a particular environment cannot support all the organisms born into that environment, and the organism best suited for the environment will have the best chance of survival.

18. **B** is correct because the scientist Jean-Baptiste Lamarck proposed the idea that acquired characteristics of the parent could be inherited by the offspring.

    A is not correct because Charles Darwin believed in a theory called "survival of the fittest," where the most "fit" organisms in an environment survive to populate the future generations in the ecosystem.

C is not correct because August Weismann was a scientist who set up experiments to disprove the evolution ideas of acquired characteristics being passed onto offspring proposed by Lamarck.

D is not correct because Hugo Vries is a scientist who first formulated the idea of mutations in an organism.

19. **C** is correct because *A. afarensis* appeared on this planet first about 3.5 million years ago.

    A is not correct because *H. sapiens* appeared on this planet about 50,000 years ago.

    B is not correct because *H. ergaster* appeared on this planet about 1.5 million years ago.

    D is not correct because *H. neanderthalensis* appeared on this planet about 80,000 years ago.

Short Response

20. The problem states that a tree contained 96 grams of carbon-14 at the time of its death and now contains 24 grams of carbon-14. It is known by scientists that half of the total grams of carbon-14 will change during each of its half-lives. Because the half-life of carbon-14 is 5,730 years and to go from 96 grams to 24 grams will take a total of two half-lives, the age of the tree would be calculated as two half-lives multiplied by 5,730 years (for each half-life), which yields a total age of 11,460 years.

21. According to Charles Darwin, an organism that is "fit" for the environment is an organism that can obtain enough food and shelter to survive and produce offspring that will create the future generations of that species. This condition can lead to the creation of a new species when a particular group of organisms in a species develop very different characteristics than other members of the same species, possibly by mutation. If these organisms with different characteristics are more "fit" for the environment than members of the same species with the original traits, and these differences eventually become so pronounced that it does not allow these two groups to mate, a new species has been created.

22. Individuals are born with a certain set of genes that allow them to survive in their environment and reproduce or not. The individual organisms cannot change this condition. A population is a group of these organisms. Some members of the population will have genetic make-ups that allow them to survive and reproduce and some members of the population will not have the genes necessary to accomplish this task. This will cause the population, not the individual organism, to change over time in response to the demands of the environment.

Ecology

There are millions of different types of life on this planet interconnected in what is sometimes called the web of life. Sometimes the connection between two or more organisms is easily seen, others times the connection is more difficult to witness. But what is important to know is that no organism, including humans, lives on this planet in isolation. **Ecology** is the study of living things and their relationship to other living things and their environment.

Lesson 6-1: How Is Life Connected?

An **organism** is usually viewed as the basic unit in the study of ecology. Each living organism found on this planet is a member of a particular species. A **species** is normally defined as a group of organisms that can mate and produce fertile offspring. All members of a species that live in a specific location create what is called a **population**. An example of a population would be a herd of bison on a specific prairie. The physical location that a group of particular organisms are best able to live is called their **habitat** or home. In fact, the word ecology means the study of home. When you gather all the plants, animals, protists, bacteria, and fungi populations that are interacting in a given location, they are called a **community**. All the living communities, or **biotic factors,** living in relationship with their **abiotic**, or nonliving environment, are called the **ecosystem**.

The world is broken up into major sections based largely on climate and the vegetation contained in the given area. These large sections of the planet are called **biomes**. The entire surface of the Earth that harbors the land, air, and water necessary for life is called the **biosphere**.

Practice Section 6-1

1. The study of the relationship between living things and the environment is called

   a) evolution b) paleontology c) ecology d) cytology

2. Which of the following would be considered a population?

   a) all the insects in the world

   b) all the birds in North America

   c) all the bullfrogs in a pond

   d) all the trees in a state park

3. Which of the following is considered to be an abiotic factor in the environment?

   a) water b) trees c) plants d) animals

4. All the living organisms and the nonliving environment in a particular location make up what is called a(n)

   a) population b) species c) biotic factors d) community

5. The physical location an organism lives is sometimes referred to as its

   a) gene pool b) biosphere c) habitat d) population

6. The portion of the earth's surface that supports life is known as

   a) the biome b) the atmosphere c) the biosphere d) the strata

7. What is a *species*?
8. What is the difference between biotic and abiotic factors?
9. What is the difference between a community and a population in an ecosystem?

Lesson 6-2: The Movement of Energy in the Environment

All life processes on Earth require the use of energy. An understanding of the flow of energy through the living and nonliving parts of the planet is one of the most important principles in comprehending ecology. Energy enters all the ecosystems on Earth from the sun. The sun must continue to supply a steady source of energy to these systems because energy cannot be recycled. When an organism uses energy, most of it becomes unusable by any other organism. The organisms on Earth that gather the energy from the sun are called the **producers**. These organisms in essence "produce" useable energy for the rest of the organisms of the ecosystem. Producers, or **autotrophs**, are able to take inorganic molecules from the environment, and by using the sun as energy, chemically create organic molecules or food. This means that even though Earth is bathed in huge amounts of sunlight each day, it is only these producers that are able to harness some of this energy for use within the living ecosystem. The chemical process where producers capture this energy is called photosynthesis. Plants, algae, and some photosynthetic bacteria are the most abundant organisms on Earth that fill the role of producer. The amount of sunlight they trap is very small compared to the amount that strikes Earth, but it is enough to sustain the biosphere.

Some organisms found on this planet are not able to make their own food. They must look for and consume organic molecules that have already been created by one of the species of producers. All organisms that must look for and then feed on organic molecules found in the environment are called **heterotrophs**. **First level consumers** are the name given to the group of heterotrophs that feed on producers for energy. First level consumers are only able to use a small fraction of the energy present inside these producers, just like the producers were only able to use a fraction of energy from the sun. Some examples of first level consumers would be herbivores such as cows, horses, deer, and rabbits.

Second level consumers (carnivores) are heterotrophic organisms that feed on first level consumers. Second level consumers obtain even a smaller fraction of that original energy from the sun that is now present in the tissue of the first level consumer they are feeding upon. This is why the energy present in an ecosystem is sometimes referred to as a pyramid. The bottom of a pyramid is the largest and would represent the producers, which are always the most abundant in the ecosystem and hold the greatest amount of stored energy from the sun. As you move up the pyramid the structure gets smaller. As you move up the ecosystem from first level consumer to second level consumer and so on, the number of these organisms in the environment decreases and the amount of trapped available energy within them also decreases.

Another type of heterotroph is called a **decomposer**. Decomposers feed on and break down organic matter in the environment such as dead plant and animal material. The kingdom of Bacteria and Fungi contain most of the species of decomposers found on the planet.

Sometimes heterotrophs can be further divided into groups according to what type of organic materials they seek for food in their environment. Heterotrophs that feed only on animal material are called **carnivores**, while heterotrophs that feed only on plant material are called **herbivores**. Some organisms feed on animal and plant material and are called **omnivores**.

Energy Pyramid

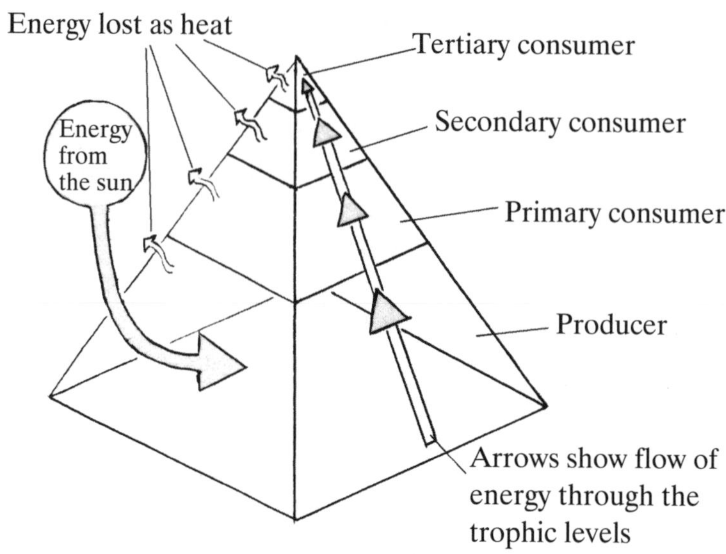

As this energy from the sun is being used by each group of organisms within the ecosystem, one could focus on each organism as it passes energy to the next group. A **food chain** is a focus on a particular pathway through the ecosystem where one organism obtains energy from the next organism by feeding on it, and so on.

This diagram represents a particular food chain that could exist in an ecosystem.

ALGAE → MINNOW → BASS → BEAR

••••••

Looking at this food chain, it would be true within that particular ecosystem that the quantity of algae would be the greatest and the number of bears would be the fewest. It would also be true that other organisms besides minnow would feed on algae, just like it would be true that other organisms besides bass would feed on minnow and so on. If you include all of the different organisms within an ecosystem that feed and obtain energy from the other organisms existing within that ecosystem, you have what is now called a food web. A food web is a collection of many food chains put together that gives a more realistic picture of the total energy flow and organisms that feed on one another within the ecosystem.

Each step in the food chain or food web is called a **trophic level.** The producers in the ecosystem would be the lowest or first trophic level and this last organism that feeds on another organism would be the highest trophic level.

Practice Section 6-2

1. An organism that cannot synthesize its own organic molecules is called a
   a) producer b) autotroph c) heterotroph d) A and B
2. Which of the following would not be considered an autotroph?
   a) plant c) photosynthetic bacteria
   b) algae d) mushroom
3. Organisms that feed on producers in the ecosystem are called
   a) first level consumers c) autotrophs
   b) second level consumers d) none of these
4. In energy pyramids, which of the following trophic levels contain the largest amounts of useable energy?
   a) producers c) second level consumers
   b) first level consumers d) third level consumers
5. Which of the following would be an example of a food chain?
   a) water → plant → sun → light energy
   b) bear → forest → cubs → den
   c) fly → spider → bird → cat
   d) soil → minerals → nutrients → plant
6. Which of the following organisms feeds on animal and plant material?
   a) carnivores b) herbivores c) omnivores d) autotrophs
7. What is the primary energy source for life on Earth?
8. What is a food chain?
9. Why is the energy present in a food web sometimes represented as a "pyramid"?

Lesson 6-3: The Movement of Materials in the Environment

As evident in the previous section, the movement of energy throughout the ecosystem is critical to all life. The availability of water, minerals, and elements such as carbon and nitrogen, are just as important as energy in the maintenance of life on this planet. These materials, unlike energy, are constantly recycled throughout the ecosystem. These materials must also be in the correct chemical form and amounts to be useful to life. The following sections will look more closely at some of the most important materials and how they are made available to living organisms.

Nitrogen Cycle

Nitrogen is an important element in living things because it is needed to build amino acids that, in turn, create proteins. Proteins are the building materials of life. About 80% of the atmosphere is made up of nitrogen gas. This large quantity of nitrogen is only useable, for energy, by certain species of bacteria that are able to convert nitrogen gas into a solid form of nitrogen called ammonia and then eventually another form of nitrogen compounds called nitrates. Plants then use the nitrates in the soil to make proteins. Animals eat plants and use the proteins in plants to make their proteins. When plants and animals die, decomposers return the nitrogen in their bodies back to the soil. Another group of bacteria can bring some of the nitrates in the soil back into the atmospheric form of nitrogen in the air in a process called **denitrification**, completing the cycle.

Humans also add many millions of tons of synthetic nitrates to the soil every year in the form of fertilizer to grow plants.

Carbon Cycle

Carbon is the structural backbone of all organic materials. Carbon is cycled by organisms in the environment through the processes of photosynthesis and respiration. Carbon is removed from molecules of carbon dioxide gas and fixed into food molecules during the process of photosynthesis. Carbon is returned to the gas form when it is released during the process of respiration as carbon dioxide. The combustion of organic molecules for the use of fuels also releases large amount of carbon in the form of carbon dioxide on this planet. The decomposition of dead organic matter can also release carbon dioxide.

Humans also release huge amounts of carbon into the atmosphere in the form of carbon dioxide every year by burning fossil fuels for energy and burning forests for open land. Some scientists believe that this large amount of carbon dioxide gas released by human activities every year could be trapping large amounts of heat in the atmosphere. This trapped heat could be slowly raising Earth's average temperature. This condition of excess carbon dioxide gas in the atmosphere raising the temperature of the planet is called the **greenhouse effect**.

Water Cycle

Water is an essential compound to all living organisms. Water can be found in the environment in lakes, rivers, oceans, as ice or snow, underground, and in the atmosphere. When the sun heats liquid water on Earth, it changes to water vapor in the process of **evaporation**. Large amounts of liquid water also are returned to the atmosphere through the process of **transpiration**. Transpiration is when water enters a plant in the liquid form and is released in vapor form during the processes of photosynthesis and respiration. The gas form of water in the atmosphere eventually cools into a liquid in the process of **condensation**. This liquid form of water falls from the sky as rain, snow, or ice. Most of this water falls into oceans or lakes found throughout world. Some water does fall on land and produces lakes, streams, rivers, ponds, and underground sources of water.

Practice Section 6-3

1. Plants are called producers because they
   a) trap light energy to help synthesize organic molecules
   b) produce carbon dioxide from oxygen
   c) produce oxygen from the process of photosynthesis
   d) none of the above
2. Which of the following is not correct concerning the nitrogen cycle?
   a) nitrogen is an element needed to produce amino acids
   b) atmospheric nitrogen is converted to ammonia by certain bacteria
   c) denitrification is when atmospheric nitrogen is changed into ammonia
   d) plants use nitrates in the soil to make proteins
3. The burning of fossil fuels releases large amounts of
   a) iron into the soil
   b) carbon into the atmosphere
   c) oxygen into the atmosphere
   d) sulfur into the soil
4. Which organisms are responsible for removing carbon from the atmosphere and fixing this carbon into organic molecules?
   a) autotrophs b) plants c) producers d) all of these
5. Transpiration is when
   a) atmospheric nitrogen is chemically fixed as ammonia
   b) ice is changed directly into water vapor
   c) liquid water that has entered a plant leaves as water vapor
   d) water vapor is changed into liquid water
6. An idea that excess carbon dioxide that is released into the atmosphere can raise the temperature of the planet is called
   a) denitrification c) the depletion of ozone
   b) the greenhouse effect d) summer solstice
7. What is an "ecological" cycle?
8. What is the importance of the element nitrogen in the environment?
9. What are the stages of the water cycle?

Lesson 6-4: Organism Interaction

Remember, the definition of ecology is the interaction of organisms with one another and the environment. These interactions among members of an ecosystem are very important factors in understanding the ecosystem.

Niche

A **niche** is the role of an organism in a community. A niche will describe things such as the food the organism eats, the place an organism sleeps, and how an organism reporduces and what it needs to do so. A niche takes into account all the biotic and abiotic factors that are part of the organism in the environment.

Example

A squirrel occupies a certain niche in a community.

Niche: A squirrel feeds on acorns, makes its nest out of dead leaves, and creates this nest on branches of certain widths at the top of trees.

Although other animals may live on or near that tree, the food requirements and shelter requirements are not the same as the squirrel's.

• • • • • •

An important concept to understand is that two organisms cannot occupy the same niche at the same time. If two organisms do occupy the same niche, competition results and only one organism will eventually remain in that particular niche. The other organism must then be able to move to a different niche in the environment or perish.

As Darwin noted in his theory of evolution, organisms are always competing for available food and shelter within an environment. **Competition** will happen any time organisms require the same food and/or shelters but there is not enough food and/or shelter for all organisms present in the ecosystem. For example, a fish-eating eagle will compete with other fish-eating eagles for food when the fish supply in a region is low. Two small plants growing close to one another will compete for light on the forest floor if the amount of light reaching the ground is limited to one spot.

Other forms of organism interaction can occur in an ecosystem besides competition. It is also true that sometimes an organism will benefit from something another organism does within an ecosystem. **Symbiosis** is an interaction among two species within an environment where at least one species obtains a biological advantage because of this interaction. There are three basic types of symbiotic relationships witnessed in any ecosystem:

1. **Mutualism** is an interaction among two species where both species benefit from the relationship. The bacteria that live in the large intestines

of humans are an example of this type of relationship. These bacteria receive food and shelter within the wall of the intestines, while at the same time they produce a vitamin needed by humans.

2. **Commensalism** is an interaction between two species where only one species benefits from the interaction. The species that does not benefit from this interaction is unaffected. An example of this type of relationship would be the remora fish that swims underneath the belly of a shark. The remora feeds on scraps of food not eaten by the shark during feeding, while the shark is unaffected by the presence of the remora fish.
3. **Parasitism** is an interaction between two species of organisms where one organism obtains a benefit and the other organism is harmed by the relationship. A dog flea is an example of parasitism. The flea feeds on the blood of the dog and obtains shelter in the hairs of the dog, while the dog is harmed by the loss of blood and possible infections or diseases that can result from the feeding of the flea.

Do not confuse parasitism with predation. **Predation** is when one organism kills and eats another organism for food. A lion feeding on a zebra is an example of predation. The lion would be considered the ***predator*** and the zebra would be considered the ***prey*** in this particular example of predation. This type of relationship among organisms will be looked at a bit more closely in Lesson 6-6.

Ecological Succession

Ecological succession is when one community gradually replaces another community through time. Each of these communities in question has its own dominant plant and animal species. Over some time interval, certain species will become less abundant and other species will become more abundant. Varying conditions in the environment bring about these community changes. For example, a small pond may gradually fill with the decayed bodies of aquatic plants until the pond is no longer a pond, but dry earth. On this dry earth, grasses will most likely grow first, then small bushes and shrubs, and eventually trees. Usually when many trees move into the community, the ability to grow is difficult for any other organism besides these trees because of the lack of sunlight and the competition for resources. A **climax community** is the final community or group of organisms that will exist in the ecosystem. The type of plant organism that is dominant in the ecosystem usually determines the type of climax community you have. The dominant type of plant usually is determined by the climate of the region in question. A climax community will not change into another community unless something such as a fire, earthquake, hurricane, or human intervention causes mass destruction of the climax vegetation.

Ecological succession appears in two forms: ***primary succession*** and ***secondary succession.*** Primary succession occurs on a site where no community had previously existed. This could include areas such as new volcanic islands or sand dunes. An area where primary succession begins has no soil. The first organisms that begin to appear in these areas are called ***pioneer organisms***. Primary succession starts very slowly because the physical environment must be changed to produce soil needed for the growth of most plant life.

Secondary succession occurs in an area that has been disturbed by such events as fire, severe weather, human agriculture or development, and even volcanism. The important difference from primary succession is that communities have previously existed on the site where this secondary succession takes place and soil is present. This factor means secondary succession is much more rapid than primary succession. As the amount of biomass (usually soil) increases, the number and variety of species will also increase in the community.

Practice Section 6-4

1. Which of the following would best describe the niche of a rabbit?
   a) its fur color
   b) how many offspring it produces
   c) the grass it eats in the environment
   d) the enzymes present in its blood
2. Which of the following conditions would probably result in the most competition for food?
   a) a lion and a zebra in the Savanna
   b) an elephant and an ant colony in a forest
   c) a spider and a plant in someone's backyard
   d) a deer and a rabbit in a grass field
3. Commensalism is
   a) an interaction among two species where both species benefit from the relationship
   b) an interaction among two species where one species benefits and the other species is unaffected
   c) an interaction among two species where one species benefits and the other species is harmed
   d) when two organisms do not interact at all in an environment
4. In the process of ecological succession, the final community or group of organisms that will exist in the ecosystem is called
   a) pioneer organisms
   b) climax community
   c) secondary consumers
   d) none of these

5. Which of the following statements are true about secondary succession?
   a) Secondary succession will begin in an area that has no soil.
   b) Secondary succession occurs on a site where no previous community existed.
   c) Secondary succession will begin in an area that has been disturbed by a condition, such as fire, that destroys the vegetation, but leaves the soil.
   d) Secondary succession will cause all plant life in the area to die.
6. The first organism that will appear in an area experiencing primary succession is called a(n)
   a) pioneer organism
   b) secondary consumer
   c) parasites
   d) heterotroph
7. Ecologists say that all organisms have a niche within the environment. What does this mean?
8. What is the difference between the organism interactions of mutualism and commensalism? Give an example of each type.
9. What is a climax community? What usually determines the nature of the climax community?

Lesson 6-5: Biomes of Earth

Parts of the planet are heated differently depending on latitude, if the area in question is close to large bodies of water, and the elevation of the physical environment. The climax vegetation of a climax community is determined by the climate of the particular region. The climate of a region is determined by the amount of sunlight it receives (temperature) and the amount of useable water present. If you look at the geographic breakdown of the planet, large areas of the world have the same climax communities. These large areas on Earth that have the same climax vegetation because of climate are called **biomes**.

As mentioned earlier, the two major influences on the type of biome found in a particular geographic area are altitude and latitude. In fact, in many cases, moving higher in elevation (altitude) on a mountain has the same effect on the type of biome you may encounter as moving north or south from the equator. Descriptions of major biomes found on Earth follow.

Tundra is found south of the permanently frozen polar areas and near the artic regions of Earth. The annual precipitation is usually less than 250 mm, and liquid water is usually not available for most living organisms. The tundra contains no trees and the ground below a few inches from the surface is permanently frozen because of the cold temperature. There is very little rainfall and the summers are short. Mosses, short grasses, and lichens are the dominant plants, while musk oxen, wolves, hares, and caribou are some of the animals able to call this place home.

Taigas are the evergreen coniferous forests that exist in a band just below the tundra in North America, Europe, and Asia. The taigas also have short, cool, summers and long, dry, cold winters. The precipitation is only about 300–500 mm annually. The ground is not permanently frozen, which permits the growth of trees. The forest trees are typically various species of firs, spruces, and some birch in the southern parts of the biome. Animals such as moose, black bears, deer, and wolves can be found in this biome.

Temperate deciduous forests contain trees that shed their leaves in the winter. Precipitation is usually anywhere between 600 and 2,500 mm annually. The winters are much milder than the taiga biome. Trees suc as oak, maple, elm, beech, and birch are common in these biomes. The understories of these forests usually contain many types of shrubs. Summers are usually of equal length as winters. The soil usually is very rich and contains a thick humus layer. Animals such as foxes, pumas, squirrels, deer, and raccoons can be found in this biome.

Tropical rainforests contain many forms of plant and animal life. Tropical rainforests are found close to the equator. In the area near the equator, temperatures, as well as precipitation, are fairly constant all year. Precipitation is around 2,000–5,000 mm annually Although it does not change much during the seasons, the climate of this region will experience wet and dry spells. The soil of most of the rainforest is very low in nutrients, partly because of the high rainfall amounts. The numbers of insects and animal species is tremendous and far greater than almost all other biomes on Earth.

Grasslands are large areas of land dominated by grasses of various species but very few trees because of the low rainfall amounts. These biomes usually receive about 250–600 mm of precipitation, mostly during the summer months. The soil in these biomes is very fertile and the humus layer can be up to a meter thick. Grasslands in North America are called prairie, in Europe and Asia they are called steppe, and in South America they are called pampa. Grazing animals such as zebra, antelope, wildebeests, bison, gophers, and gazelles are common in this biome.

Deserts are areas on Earth that experience very little rainfall and have very hot days and very cold nights. Precipitation in these biomes is usually less than 250 mm annually. Droughts, or long periods with almost no rain, are common. Because of the lack of rainfall, trees are rare, and in some desert biomes even grass vegetation is not common. The soil for the most part is very poor in nutrients. Cacti and small shrubs and animals such as lizards, small rodents, and jackrabbits are sometimes common in these biomes.

Savannas are a type of grassland biome located between deserts and tropical rainforests in the warmer climate regions of the world. Precipitation is usually around 900–1500 mm annually. Droughts are very common in these biomes, and most plant life will die or become dormant until the rain arrives. Soils are usually not rich because of the rainy season wash-off. Some trees are present in this biomes, but there are mostly species of grass.

Aquatic biomes are the waters of Earth. Water covers more than 75% of Earth's surface. All this water can divided into two major groups: marine, or saltwater, biomes and freshwater biomes.

Ocean biomes contain huge bodies of salt water that have great influences over the climates of the entire biosphere. The animal and plant life found within the ocean varies depending on the depth of the water, its proximity to land, and the temperature of the water. Because the amount of space that oceans take up is so large, ecologists divide the ocean into sections for study.

Zones in the Marine Biome

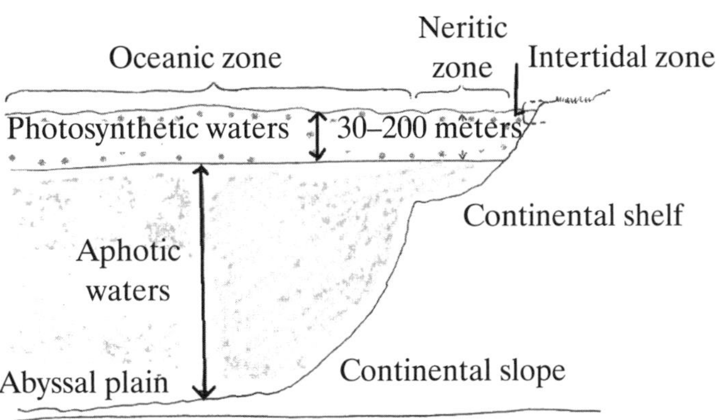

The ocean is divided according to water depth into three areas called the **intertidal zone**, the **neritic zone**, and the **oceanic zone**.

The intertidal zone is the shoreline area of the ocean. This part of the ocean is very shallow and will be exposed to air at varying times during the day at low tide. Organisms such as mussels and crabs will inhabit this area because they are able to live out of water for short periods of time.

The neritic zone is the region of the ocean just off shore and before the deep sea. The depth of water in the neritic zone can range from 30 to 200 meters. Most fish as well as sea mammals live in this part of the ocean.

The oceanic zone is the open sea and the depth of water in this region is usually hundreds of meters deep. Very little sea life is usually found in this region except for near the surface where there are many photosynthetic algae and the small creatures that live on them. Ocean producers such as algae synthesize more organic food molecules than any other creature on Earth.

Freshwater biomes include the lakes, rivers, ponds, and streams of the earth. The waters of this biome have low salt concentrations. Usually the amount of dissolved oxygen in the water, the temperature of the water, and the amount of nutrients in the water, determine what kinds of organisms and their quantities found in areas of this biome. Organisms like trout, sunfish, bass and perch can be very plentiful. Many birds and land animals use these freshwater biomes as nesting and feeding sites.

Estuaries are areas where freshwater, such as a stream, runs into salt water. Water in an estuary is a mix of fresh and salt. The waters of an estuary are very rich in nutrients and are important breeding grounds for many organisms.

The Biomes of the World

Tundra	**Tropical Rainforests**
Grasslands	**Deserts**
Taigas	**Savannas**
Temperate Deciduous Forests	**Ocean/Freshwater**

••••••

Practice Section 6-5

1. What characteristic is usually used as the major identifying factor of a particular biome?
   a) latitude
   b) longitude
   c) rock formations
   d) climax plant life
2. Which of the following biomes receives the highest annual rainfall?
   a) taigas
   b) deserts
   c) grasslands
   d) tropical rain forests
3. Which of the following biomes contains many trees that lose their leaves in the colder months of the year?
   a) tropical rain forest
   b) grasslands
   c) temperate deciduous forest
   d) tundra
4. Which climax vegetation indicates a taiga biome?
   a) grasses
   b) mosses
   c) deciduous trees
   d) coniferous trees

5. A biome is usually determined by
   a) latitude
   b) altitude
   c) amount of rainfall
   d) all of these
6. How does a marine biome differ from a freshwater biome?
   a) amount of sunlight that strikes the surface of the water
   b) amount of salt dissolved in the water
   c) amount of algae that is found in the water
   d) the temperature of the water
7. What is a biome? Give an example of one.
8. What are the zones of an ocean biome?
9. What biomes are found in the United States?

Terms From Chapter 6	
abiotic factors	greenhouse effect
autotrophy	habitat
biome	heterotroph
biosphere	mutualism
biotic factors	niche
commensalisms	parasitism
community	population
competition	predation
condensation	predator
denitrification	prey
ecology	producer
ecosystem	species
evaporation	symbiosis
food chain	transpiration
food web	trophic level

Chapter 6: Exam

Matching Column for Biomes

Match the biome with the best description of that biome. You may use an answer choice more than once or not at all.

Biome		Description
1. tundra	a.	This area receives very little rainfall, and usually experiences very hot days and cold nights.
2. grassland	b.	This area is usually dominated with tree species that shed their leaves in the winter months.
3. tropical rainforest	c.	This area has permanently frozen soil that does not allow for the growth of large plants.
4. savanna	d.	This area has few trees, very fertile soil, and usually many species of grasses. The rainfall amounts are low and the rain is more abundant during the summer months.
5. desert	e.	This biome is usually located near the equator. Rainfall amounts are very high, vegetation is dense, and soil quality is poor.
6. temperate deciduous forest	f.	This area is dominated by conifer trees and short summers.
7. taigas	g.	A type of grassland biome that experiences rainy seasons and long periods of drought.

Matching Column for Symbiotic Relationships

Match the symbiotic relationship with the best description of that relationship. You may use an answer choice more than once or not at all.

Symbiotic relationship		Characteristic
8. mutualism	a.	An interaction among organisms in which one organism benefits from the relationship and the other organism is harmed.
9. commensalism	b.	An interaction among organisms in which both organisms benefit from the relationship.
10. parasitism	c.	An interaction in which one organism kills and eats the other organism.
11. predation	d.	An interaction among organisms in which one organism benefits from the relationship and the other organism is unaffected.

Multiple Choice

12. What is the ecological name given to organisms that feed on producers?
    a) autotrophs c) second level consumers
    b) first level consumers d) predator

13. Which of the following is an example of a food chain?
    a) dirt → minerals → light → plants
    b) water → hydrogen → oxygen → decomposition
    c) algae → minnow → bass → raccoon
    d) organism → species → community → ecosystem

14. All of the following are found in the biosphere except the
    a) oceans b) atmosphere c) forests d) moon

15. What percent of the atmosphere is nitrogen gas?
    a) 20% b) 40% c) 60% d) 80%

16. When released in large quantities into the atmosphere, what gas do scientists believe may be responsible for the greenhouse effect?
    a) oxygen b) carbon dioxide c) nitrogen d) sulfur dioxide

17. A tick feeding on the blood of a house cat is an example of
    a) mutualism c) predator-prey
    b) commensalisms d) parasitism

18. Which biome on the planet experiences relatively small precipitation amounts and the dominant vegetation are coniferous trees?
    a) tropical rain forest c) taiga
    b) savanna d) tundra

19. What section of the marine biome would you find the largest concentration of sea mammals and fish?
    a) neritic zone b) oceanic zone c) deep sea d) intertidal zone

Short Response

20. Explain why each trophic level in a food chain contains less energy than the level below it.
21. Nutrients are recycled in the ecosystem, but energy is not. Explain.
22. Explain the difference between the niche of an organism and its habitat.

Answer Key

Answers Explained Section 6-1

1. **C** is the correct answer because the study of the relationship between living things and their environment is called ecology.

   A is not correct because evolution is the study of how life on this planet has changed from ancient forms to present day forms.

   B is not correct because paleontology is the study of the remains of ancient life on this planet.

   D is not correct because cytology is the study of the cell and its functions and structures.

2. **C** is the correct answer because a population would be considered all the bullfrogs in a pond. A population is all the members of a specific species in a specific geographic area.

   A is not correct because all the insects in the world would not be a population because this would include all the species of insects in a very general geographic area.

   B is not correct because all the birds in North America would not be a population because this would include all the species of birds in a very general geographic area.

   D is not correct because all the trees in a state park would not be a population because it would include all the species of trees in this area.

3. **A** is the correct answer because the abiotic factor or nonliving substance in the list of items is water.

   B, **C**, and **D** are not correct because trees, plants, and animals are all living (biotic) factors in the environment.

4. **D** is the correct answer because all the living organisms and the nonliving environment in a particular location are called a community.

   A is not correct because population is a group of organisms of the same species in a specific geographic location.

   B is not correct because species are members of a group that can mate and produce fertile offspring.

   C is not correct because biotic factors are the living components of an ecosystem.

5. **C** is the correct answer because the physical location at which an organism lives is called its habitat.

   A is not correct because gene pool is all the genes present for the members of a specific population.

B is not correct because biosphere is any area on this planet where life is found.

D is not correct because population is a group of organisms of the same species in a specific geographic location.

6. **C** is the correct answer because the portion of Earth's surface that supports life is known as the biosphere.

   A is not correct because biome is a geographic location on the planet with a specific climate and vegetation.

   B is not correct because atmosphere is the mixture of gases that surrounds the surface of the planet.

   D is not correct because strata is the name of rock found in layers underneath Earth's surface.

7. A species is the most basic unit of biological classification. A species is normally defined as a group of organisms capable of interbreeding and producing fertile offspring.

8. Biotic factors are the living things (plants, animals, bacteria, and so on) that make up the ecosystem. Abiotic factors are the non-living elements (air currents, water, soil, and so on) that make-up the ecosystem.

9. A population is group of organisms from the same species, living in the same location, during the same period of time. A community is a group of populations living and interacting together.

Answers Explained Section 6-2

1. **C** is the correct answer because an organism that cannot synthesize its own organic molecules is called a heterotroph.

   A and **B** are not correct because producers and autotrophs are organisms that can synthesize their own organic molecules.

2. **D** is the correct answer because a mushroom is not considered to be an autotroph. Mushrooms are types of fungi that are called absorptive heterotrophs.

   A, **B**, and **C** are not correct because plants, algae, and photosynthetic bacteria are all autotrophic organisms.

3. **A** is the correct answer because organisms that feed on producers are called first level consumers.

   B is not correct because second level consumers feed on first level consumers in a food chain.

   C is not correct because autotrophs are considered to be producers. Producers and autotrophs make their own organic molecules for food.

4. **A** is the correct answer because in an ecological energy pyramid, the trophic level that contains organisms with the most useable energy is the producer level. There are more producers than consumers in an ecological system.

   B is not correct because the trophic level that contains first level consumers holds the second most useable energy.

   C is not correct because the trophic level that contains the second level consumers holds less energy than both the producer and first level consumer trophic groups.

   D is not correct because the third level consumer trophic group contains the least amount of energy of all the groups present.

5. **C** is the correct answer because an example of a possible food chain from the list given is: fly → spider → bird → cat. A food chain shows the energy flow and feeding relationships of a group of organisms in a particular ecosystem.

   A, **B**, and **D** are not correct because these other answers are items in an ecosystem that could be grouped together but do not show energy or feeding relationships.

6. **C** is the correct answer because omnivores feed on both plant and animal material.

   A is not correct because carnivores are organisms that feed only on animal matter.

   B is not correct because herbivores are organisms that feed only on plant matter.

   D is not correct because autotrophs are organisms that make their own food.

7. The primary source of energy for life on Earth is the Sun. The sunlight is trapped and used by autotrophic organisms to produce organic food molecules. These organic food molecules are then consumed by other organisms.

8. The food chain is the connection between predator and prey from autotrophic organisms to the top consumers. Food energy flows from the beginning of the food chain (autotrophs) to the end of the food chain (consumers). Many food chains connected together make-up a food web.

9. The reason the energy present in a food web is sometimes represented as a pyramid is because the base of this pyramid (the largest section), represents the most abundant group the autotrophs. The autotrophic organisms contain the most trapped energy from the sun. As you move up the pyramid, the primary consumers are less abundant then the autotrophs below them, and thus have less available energy. This trend continues to the top of the pyramid.

Answers Explained Section 6-3

1. **A** is the correct answer because plants are called producers because they trap light energy to help synthesize organic molecules. These plants "produce" food from this light energy.

   B is not correct because carbon dioxide is not produced from oxygen.

   C is not correct because oxygen is produced from the process of photosynthesis, but this is not why plants are called producers.

2. **C** is the correct answer because denitrification is not when atmospheric nitrogen is changed into ammonia. Denitrification is when nitrates in the soil are brought back into the atmospheric form of nitrogen in the air.

   A, **B**, and **D** are not correct because all of the following are correct reactions in the nitrogen cycle: nitrogen is an element needed to make amino acids and proteins, atmospheric nitrogen is converted to ammonia, and plants use nitrates in the soil to make proteins.

3. **B** is the correct answer because fossil fuels release large amounts of carbon into the atmosphere.

   A, **C**, and **D** are not correct because large amounts of these other elements, iron, oxygen, and sulfur, are not released in amounts as large as carbon during the burning of fossil fuels.

4. **D** is the correct answer because autotrophs, plants, and producers are all organisms that remove carbon from the air during the process of photosynthesis.

5. **C** the correct answer because transpiration is when liquid water that has entered a plant leaves as water vapor. Most of this water leaves the plant from the openings on the leaves called stomata.

   A is not correct because when atmospheric nitrogen is chemically converted to ammonia in the soil this process is called nitrogen fixation.

   B is not correct because sublimation is when ice is transformed directly into water vapor.

   D is not correct because condensation is when water vapor is changed into liquid water.

6. **B** is the correct answer because the greenhouse effect is the name given to the idea that if large amounts of carbon dioxide are released into the atmosphere, this can, in turn, raise temperatures globally. This idea is based on the premise that carbon dioxide has a tendency to hold heat.

   A is not correct because denitrification is when nitrates in the soil are changed back into the atmospheric form of nitrogen.

C is not correct because depletion of ozone is believed to be caused by large amounts of aerosol chemicals that are released into the atmosphere causing the breakdown of ozone molecules in the upper atmosphere.

D is not correct because summer solstice is the name given to the time of year when Earth's northern hemisphere is tilted at its greatest angle toward the sun.

7. An ecological cycle is when a particular material, organic or inorganic, makes its' way repeatedly through the environment. Following this material through the cycle enables one to understand the interactions that occur with the particular material and the things it comes across in the environment.

8. Nitrogen is the basic element needed to create amino acids. Amino acids are the subunits of all proteins. Proteins are the building blocks of life. This means that all living organisms need nitrogen to survive.

9. The water cycle is the movement of molecules of H_2O throughout the environment. When liquid water, on and in the ground, is heated it evaporates into water vapor in the air. The plant process of transpiration also puts huge amounts of water vapor into the air also. When water vapor in the air is cooled it condenses back into liquid water and falls to the earth as rain or snow. Then process repeats itself continually.

Answers Explained Section 6-4

1. **C** is the correct answer because the niche of an organism is its role in the environment. A niche describes things such as the food eaten by the organism and the place an organism sleeps.

   A is not correct because the fur color of a rabbit describes its phenotype based on its genetic inheritance.

   B is not correct because the number of offspring the organism will produce is the reproduction rate.

   D is not correct because the enzymes present in the blood of the organisms are based on the species of organism and the genotype of that organism.

2. **D** is the correct answer because a deer and a rabbit in a grass field would result in the most competition, because both a deer and a rabbit feed on the same substance in the environment.

   A is not correct because a lion and a zebra do not feed on the same substances, and so there would be no competition for food. A lion and a zebra could be considered predator and prey within an ecosystem because lions will eat zebras for food.

B is not correct because an elephant and the organisms present in an ant colony really would not be in competition for the same food or sleeping space.

C is not correct because a spider and a plant are organisms that would not be competition for food within the environment.

3. **B** is the correct answer because commensalism is a type of symbiotic relationship where one organism of a species benefits from the relationship and the other organism in a different species is unaffected.

   A is not correct because an interaction among two species where both species benefit from the relationship is called mutualism.

   C is not correct because an interaction among two species where one species benefits and the others species is harmed is called parasitism.

   D is not correct because when two organisms from different species do not interact in the environment there is no symbiotic relationship between them.

4. **B** is the correct answer because the final community of organisms in ecological succession is called a climax community.

   A is not correct because pioneer organisms are the first organisms to appear in an area experiencing primary succession.

   C is not correct because secondary consumers are the organisms in a food chain that feed on the primary consumers.

5. **C** is the correct answer because secondary succession begins in an area that has been disturbed and loses the vegetation and animal life of the region but the soil layer remains.

   A is not correct because primary not secondary succession begins in an area that has no soil.

   B is not correct because primary not secondary succession begins in an area where no previous community existed.

   D a natural or manmade disturbance, such as fire, can cause all the plant life in area to be killed.

6. **A** is the correct answer because the first organism that appears in an area experiencing primary succession is called a pioneer organism.

   B is not correct because a secondary consumer is the organism that feeds on the primary consumers in a food chain.

   C is not correct because parasites are organisms that benefit from other organisms in the environment and cause harm to these organisms in the process.

D is not correct because heterotrophs are organisms that are not able to synthesize their own organic molecules for food.

7. A niche is the role a particular organism has within a particular community of the environment. A niche describes the nutrition of the organism, where it lives, how it breeds, how it interacts with all the living and non-living things it comes into contact with. For example an oak tree growing in a deciduous forest in the northeast United States occupies a particular niche. This oak tree absorbs water and certain minerals from the dirt it is growing in. This oak tree absorbs sunlight and carbon dioxide. This oak tree provided a home for many species of birds and mammals. This oak tree showers the ground each fall with acorns. Although other organisms that live in the same forest as this oak tree may do some of these things, this particular oak tree is the only one that does them all. This is what creates the niche.

8. Mutualism and commensalism are both types of symbiosis. Mutualism is an interaction between two species of organisms where both organisms benefit from the relationship. An example of mutualism would be a honeybee obtaining pollen from the flower of a plant for food. The plant also benefits, because this act by the bee enables the plant to reproduce. Commensalism is an interaction among two species of organisms where one organism benefits and the other organism is unaffected. An example of commensalism would be the barnacles found attached to the skin of certain species of whale. The barnacles obtain a home base and transport, the whale is unaffected.

9. A climax community is the final species group found in a particular ecosystem. This final species group is most often determined by the climate (temperature, rainfall amount, humidity, and so on) of a particular region.

Answers Explained Section 6-5

1. **D** is the correct answer because the climax plant life is usually the main identifying factor of a particular biome of the world.

   A is not correct because the latitude of the particular geographic area, or the distance north or south from the equator, can at times determine the possible biome that could be present in a certain region, but not always. Altitude with latitude would be important in making this determination about biome.

   B is not correct because the longitude, which is the distance east or west from the prime meridian of a particular geographic area, usually does not determine a biome of a region.

   C is not correct because the rock formations present in a particular region will not usually be a major factor in determining the particular biome of an area.

2. **D** is the correct answer because most tropical rainforests will receive higher amounts of rainfall than other biomes located on the planet.

   A and **B** are not correct because taigas and deserts usually receive moderate to small amounts of rainfall in a given year.

   C is not correct because at times during the year, grasslands can receive as much rainfall as tropical rainforests, but will also go through long periods of drought at other times of the year.

3. **C** is the correct answer because biomes called temperate deciduous forests contain many species that will lose their leaves in the winter, such as oak and maple trees.

   A is not correct because tropical rainforests do not really experience the seasons like other biomes on the planet. The trees of a tropical rainforest will not lose their leaves at certain times of the year.

   B is not correct because grassland biomes usually have few to no trees in this geographic region of the planet.

   D is not correct because tundra is a biome of the planet that has no trees.

4. **D** is the correct answer because the taiga biome is usually dominated by coniferous or evergreen trees because of the climate of cold winters and dry summers.

   A is not correct because grass species are the climax community found in a grassland biome.

   B is not correct because mosses are the climax community found in the tundra.

   C is not correct because deciduous trees are the climax plant community found in a temperate deciduous forest.

5. **D** is the correct answer because a biome is usually determined by all of the factors (latitude, altitude, and amount of rainfall) working together in the environment.

6. **B** is the correct answer because a marine biome is contrasted to a freshwater biome by the amount of dissolved salts found in the water. In marine biomes, there is much more dissolved salt found in the water than in freshwater biomes.

   A is not correct because the amount of sunlight that strikes a given surface area of water will vary in both marine and freshwater biomes.

   C is not correct because the amount of algae found in the water can be both high and low in freshwater and marine biomes.

   D is not correct because the temperature of the water can be both high and low in freshwater and marine biomes.

7. A biome is a particular geographic region on the planet that has similar climate conditions. The plant life existing in the region is the chief factor in determining the type of biome present. An example of a biome would be a tropical forest. A tropical forest biome is characterized by tall trees in a region of year-round warmth. An average of 50 to 260 inches of rain falls yearly.
8. The zone of an ocean biome includes: the oceanic zone (all the water over the open ocean), the neritic zone (all the water above the continental shelves), and the intertidal zone (the area of ocean that sometimes experiences dry land).
9. The biomes found in the United States include: rainforest (Hawaii), grasslands (Texas), tundra (Alaska), deciduous forest (Maine), coniferous forest (North Dakota), and desert (Nevada).

Answers Explained Chapter 6 Exam

Matching Column

1. C	4. G	7. F	10. A
2. D	5. A	8. B	11. C
3. E	6. B	9. D	

Multiple Choice

12. **B** is the correct answer because all organisms that feed on producers are called first level consumers.

    A is not correct because autotrophs and producers are the same type of organism.

    C is not correct because second level consumers feed on first level consumers, not producers.

    D is not correct because predators are organisms that feed on other not autotrophic organisms.

13. **C** is the correct answer because this choice shows what organisms certain other organisms obtain energy from in the environment. A food chain shows a link of feeding relationships for a particular environment.

    A, **B**, and **D** show some groups of biological words, but with no specific feeding connection or shown energy transfer.

14. **D** is the correct answer because the moon is not found in the immediate area of Earth's crust. The upper crust and the surrounding atmosphere of the planet Earth are considered to be the biosphere. The moon is many miles beyond this zone.

A, B, and C are not correct because the oceans, atmosphere, and forests of the planet Earth are all part of the biosphere.

15. **D** is the correct answer because 80% of Earth's atmosphere is nitrogen gas.

16 **B** is the correct answer because many scientists believe that when large amounts of carbon dioxide gas are released in the atmosphere it will have a warming affect on the climate of the world.

A, **C**, and **D** are all gases found in the atmosphere, but are not believed to be linked to the greenhouse effect proposed by some scientists.

17. **D** is the correct answer because when a tick feeds on the blood of a house cat, this is an example of parasitism. Parasitism is when one organism benefits (the tick) and the other organism in the relationship is harmed (the cat).

A is not correct because mutualism is when both organisms in a relationship benefit from the relationship.

B is not correct because commensalism is when one organism benefits and the other organism is unaffected in a symbiotic relationship.

C is not correct because in a predator-prey relationship, one organism must kill and eat the other organism.

18. **C** is the correct answer because the taiga is the biome that experiences small amounts of precipitation and is dominated by conifer trees.

A is not correct because tropical rainforests experience large amounts of precipitation.

B is not correct because savannas have very few trees and none of them are conifers.

D is not correct because the tundra is dominated by mosses and has almost no trees.

19. **A** is the correct answer because the neritic zone that lies just off shore usually contains the highest numbers of fish and sea mammals. This area is highly concentrated with sea life because there is a relatively high concentration of food available in this zone for these organisms.

B and **C** are not correct because the oceanic zone and the deep sea have very little food available to support large numbers of larger living organisms.

D is not correct because the intertidal zone is periodically dry during low tide, meaning many fish cannot survive in this area.

Short Response

20. A trophic level is a group of organisms in a particular environment that feed on the same energy level of organisms in the ecosystem. The first trophic level in the ecosystem contains the producers. The organisms in this level contain the most stored energy, because they obtain this energy directly from the sun. An organism that feeds on a producer can only obtain some of that stored energy during nutrition; some of the stored energy is always lost as heat before it can be transferred to the next organism in the food chain. As you continue up the trophic levels from producers to consumers, more and more of the energy that is stored is lost as heat before it gets to the next level of organisms.

21. Nutrients are recycled in the environment because all nutrients are made of matter. This is a law in science that states matter can neither be created nor destroyed—only changed in form. When nutrients are recycled in the environment they are not being destroyed, they are just being sorted with different substances. So when these nutrients are being changed in form, these forms are still useable.

    There is a law in science that states that energy can neither be created nor destroyed—only changed in form. The problem is that some forms of energy (such as heat) are not useable by many organisms. This means that all energy eventually reaches a form unusable by most living organisms in the environment.

22. The habitat is the actual physical geographic area in which the organism dwells. The niche is the role that an organism plays in the environment. A niche includes the food requirements, the shelter needed, and the environmental interactions of an organism.

7

The Human Body

The human body is a miraculous living machine. Trillions of living cells all with specific functions work together, like ants in a colony, to create a working community of life. All of the cells of the human body are grouped into specialized tissue that perform all the unique functions of the life processes. In this chapter, we will look at these different unique tasks that these cells must accomplish to keep the body alive.

Lesson 7-1: Human Nutrition

The human body is composed of an army of cells all working together. As the saying goes, an army moves on its stomach. This army of cells must be fed to work properly. Cells need energy and chemical supplies to perform actions such as growing, repairing worn-out parts of the body, maintaining homeostasis, and reproducing.

Humans are heterotrophic organisms, which means they must obtain organic compounds from the environment for nutrition to feed this army of cells. Let us look closer at how the food enters the human body, and how the body distributes this food to all its cells.

Major Parts of the Human Digestive System	
Oral Cavity	**Large Intestines**
Esophagus	**Rectum**
Stomach	**Liver**
Small Intestines	**Pancreas**

• • • • • •

Oral Cavity

Human Digestive System

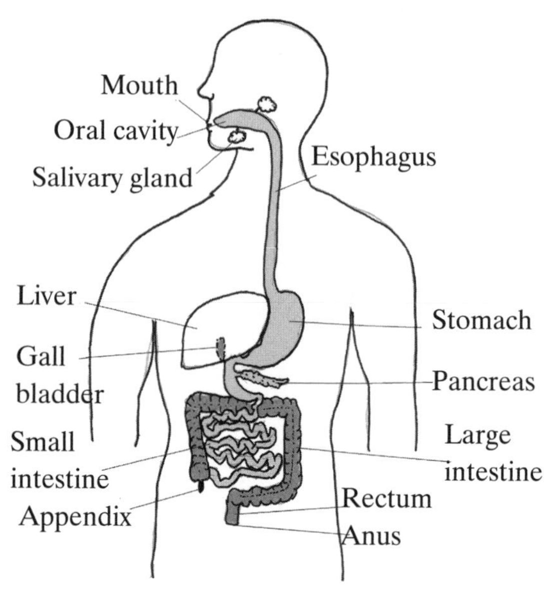

Food or organic compounds are ***ingested***, or enter the human body via the **oral cavity**, or mouth. In the mouth food is both chemically and physically digested. Teeth grind and soften the food into small pieces that will be easier to chemically digest later and also make the food easier to swallow. This grinding and softening of the food is physical digestion, which means the food is reduced in size but not changed in chemical nature. For example, if someone is chewing on a piece of steak, the teeth are making the steak pieces smaller and softer, but they are still pieces of steak. This is why the grinding of teeth against the food is called physical digestion.

But some chemical digestion does occur in the mouth. The enzyme ***ptyalin***, which is a kind of enzyme called ***amylase*** that digests carbohydrates, is released by salivary glands in the mouth. Ptyalin is an enzyme that chemically digests starch molecules into dissacharides. The food is covered in ***mucus*** before it is swallowed to lubricate its passage into the stomach from the mouth. Swallowed food then enters the tube that connects the mouth with the stomach, called the **esophagus**. At the top of the esophagus there is a small flap called the ***epiglottis***, which prevents food from entering the windpipe, or ***trachea***, as food passes by its entrance on the way to the stomach. Food moves through the esophagus by muscular contractions known as ***peristalsis***. The muscular contractions of peristalsis move food through the esophagus in a very similar fashion as the way someone would squeeze toothpaste out of the tube. It is peristalsis that moves food through this tube, not gravity. That is why, although not recommended, you could swallow food and it would enter your stomach even standing on your head. At the base of the esophagus is a muscular organ called the ***cardiac sphincter***. The cardiac sphincter opens and closes the entrance into the stomach from the esophagus. When food is in the stomach and being digested, this muscle makes sure the opening to the esophagus is closed, so that food, digestive enzymes, and acids don't work their way back up the esophagus. Sometimes, when these strong acids from the stomach do make their way up the esophagus they can burn the esophagus lining. This is a condition most people know as heartburn. It is only called heartburn because the esophagus, which is actually receiving the burning sensation, is found next to the heart.

Stomach

The stomach begins the next step of physical and chemical digestion. The **stomach** is a muscular pouch that can expand to both store and digest food. Gastric juices secreted by the stomach aid in the process of chemical protein digestion. Gastric juice contains ***hydrochloric acid*** and ***pepsinogen***. Hydrochloric acid (HCl) has two functions in the stomach. HCl is a strong acid that kills most of the bacteria swallowed with food; it also changes the chemical pepsinogen into the active form of the enzyme ***pepsin***. Pepsin is an enzyme that begins the chemical breakdown of protein in the stomach. The chemical breakdown of protein means the steak is now being reduced into simpler, new substances, such as amino acids. All these substances found in the stomach, such as water, HCl, pepsin, and chewed food, form a chunky liquid mass called ***chyme***. When the food has been stored and digested for a proper length of time, the **pyloric sphincter** muscle releases these substances into the small intestines from the stomach.

Usually the stomach lining is protected from acid erosion by a thick layer of mucus. But sometimes the strong acids in the stomach will not only digest the food present, but will also digest the lining of the stomach itself. This condition is called an ***ulcer***.

Small and Large Intestines

The tube from which food leaves the stomach is called the **small intestines**. It is called "small" not because of length but because of diameter compared to the large intestines. When food enters the small intestine, which is a six-meter coiled long tube in the lower abdomen, digestive juices from three sources are mixed with the food. Bile from the liver; pancreatic enzymes, such as ***proteases***, ***amylases***, and ***lipases*** from the pancreas; and digestive enzymes from the lining of the small intestine itself, all chemically digest food that enters. Bile breaks fat up into small globules, which are easier to chemically digest because there is more surface area exposed. This process of breaking fat into smaller globules is called **emulsification**. Bile also neutralizes the strong acids exiting the stomach so that they do not harm the lining of the small intestine. Bile contains strong alkaline (base) chemicals that accomplish this task of neutralizing the acids from the stomach. Proteases are a group of enzymes that aid in the chemical digestion of proteins, amylases are a group of enzymes that aid in the chemical digestion of carbohydrates, and lipases are a group of enzymes that aid in the chemical digestion of lipids.

If the inner walls of the small intestines are observed closely, they are not smooth, but rough with millions of tiny hairlike extensions called **villi**.

The villi increase the surface area of the small intestines, allowing much more food to be absorbed into the bloodstream than could ever be absorbed without them. Inside each of the millions of villi are small capillaries and small tubes called **lacteals**. The capillaries absorb some of the by-products of chemical digestion,

Structure of the Villus

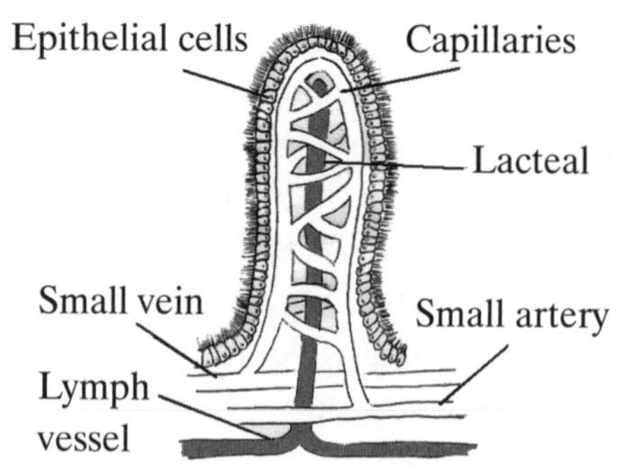

such as amino acids and simple sugars, while the lacteals absorb the subunits from the breakdown of lipids, such as glycerol and fatty acids.

Any food that is not able to be absorbed by the small intestines or has not had the proper time to be digested by the rest of the body will enter the **large intestines**. The large intestine is about one and a half meters long, which is shorter than the small intestines, but it is called "large" because the diameter of this tube is wider than that of the small intestine. Food is not digested in the large intestine. There are two important processes that take place in this tube: water absorption and vitamin production. First, most of the water in the undigested food is absorbed by the walls of the large intestine and dumped back into the bloodstream. Second, bacteria that live in a mutualistic symbiotic relationship with humans feed on some of the organic products in the intestine while simultaneously producing needed vitamins for the host.

All the leftover undigested food, or ***feces***, is stored in the lowest part of the large intestine called the **rectum**. The rectum stores this waste until it is eventually eliminated through the **anus**.

Practice Section 7-1

1. The chemical digestion of proteins begins in the
   a) oral cavity b) stomach c) small intestines d) large intestines
2. Amylase is an enzyme used to aid in the chemical digestion of
   a) proteins b) carbohydrates c) lipids d) water
3. A series of wavelike muscle contractions that bring food down the esophagus into the stomach is called
   a) pepsin c) peristalsis
   b) dehydration synthesis d) angina
4. In which of the following structures does no digestion of food occur?
   a) oral cavity b) stomach c) small intestines d) large intestines
5. Emulsification is
   a) the digestion of proteins in the stomach
   b) the physical breakdown of lipids in the small intestines
   c) the digestion of carbohydrates in the mouth
   d) the removal of undigested food from the rectum

6. Which of the following enzymes are released into the small intestines by the pancreas?
   a) amylases b) lipases c) proteases d) all of these
7. The stomach contains an enzyme called pepsin that helps in the breakdown of what macromolecule?
   a) lipids b) carbohydrates c) water d) proteins
8. A lipid is broken down into glycerol and fatty acid molecules, this is an example of a
   a) mechanical digestion c) dehydration synthesis
   b) chemical digestion d) B and C
9. What is the name of the structures found inside the lining of the small intestine that greatly increase the surface area of this tube?
   a) villi b) cilia c) flagella d) capillaries
10. What structure covers the trachea during the swallowing of food or drink to prevent these materials from entering the windpipe?
    a) larynx b) epiglottis c) pharynx d) villi
11. What type of digestion occurs in the human mouth?
12. What are the different functions of the small and large intestines in humans?
13. What is an ulcer?

Lesson 7-2: Human Transport

Food and oxygen can enter the body only in certain locations, but every cell in the body needs to receive these materials. Transport is the process that distributes these supplies to all the cells. Transport is the movement of materials into the cells and the movement of materials throughout the body of an organism. In most one-celled organisms, the transport of materials such as food, oxygen, and waste products is accomplished by diffusion across the cell membrane to or from the immediate environment. In larger multicelled organisms, such as humans, many of the cells are not in contact with the environment. These materials must be shuttled back and forth from the internal cells and the environment. Humans have a closed circulatory system that accomplishes this task. The circulatory system of humans includes the blood vessels, the blood, the lymphatic system, and the heart. We will look at each part of this system more closely.

Major Parts of the Human Circulatory System

Blood **Blood Vessels** **Heart** **Lymph**

••••••

The Heart

For the army of cells to be constantly supplied with the materials necessary for life, the blood in humans must constantly be moving materials to and from these many trillions of cells in the body. The **heart** is the pumping mechanism that keeps the blood in this constant motion. The human heart is a strong, muscular organ that contains four chambers. The upper chambers are called the **atria**, while the lower chambers are called the **ventricles**. The pathway of blood into and out of the heart is illustrated in the diagram.

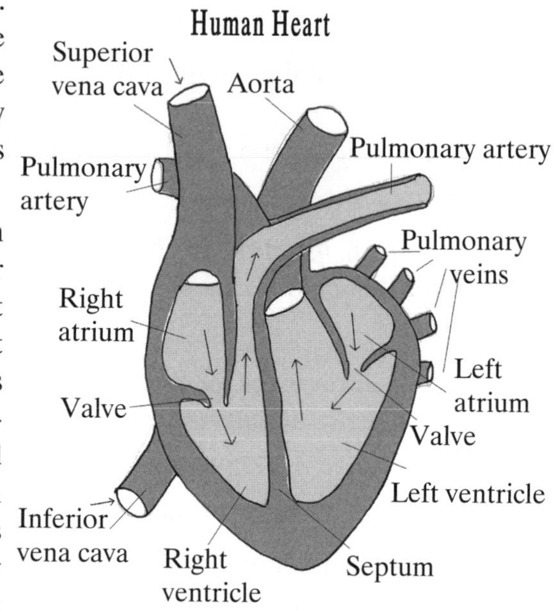

Blood that is low in oxygen returns from the upper and lower parts of the body and enters the right atrium of the heart and then the right ventricle. The right ventricle pumps the deoxygenated blood into the pulmonary arteries that bring the blood to the lungs. The red blood cells in the blood receive oxygen in the lungs and return by way of the pulmonary veins to the left atrium and then the left ventricle. The left ventricle pumps the oxygenated blood out of the heart by way of the largest artery in the body called the aorta to all the parts of the body. The blood will eventually return to the right atrium of the heart to complete the cycle.

Coronary blood vessels branch from the aorta down to the muscle tissue of the heart itself. Other blood vessels originate in the heart tissue itself and bring deoxygenated blood back mainly to the right atrium. These coronary blood vessels supply the heart muscle with food and oxygen, and also remove waste from the tissue. Even though the heart has blood flowing through it internally all the time, it is unable to obtain the materials it needs from this blood. This makes the coronary blood vessels essential for the health of the heart. A heart attack is when these coronary arteries become blocked and do not allow the heart tissue to receive food and oxygen from the blood. This causes death of that heart tissue and the chest pain associated with a heart attack.

Blood Vessels

The highway system of the body is composed of the blood vessels. Blood vessels are tubular structures that transport blood throughout the body. The three major blood vessels found in the human body are arteries, veins, and capillaries.

Arteries are thick-walled blood vessels that carry blood under high pressure from the heart outward to the extremities of the body. The blood in the arteries is usually under high pressure because the blood has recently left the pumping mechanism of the heart. The arteries contain oxygenated blood because they are carrying blood out to the body from the left side of the heart.

As the arteries bring blood to the outer extremities of the body, they become smaller and smaller in diameter. Eventually blood vessels that are the thickness of one cell branch off from the larger arteries. These very narrow blood vessels are called capillaries. **Capillaries** are the blood vessels in which most of the diffusion of material actually occurs from the blood to the surrounding tissue. The small diameter of these tubes allows the passage of oxygen and food. Capillaries also collect the waste produced from the cells of surrounding tissue and brings them back toward the heart. These capillaries gradually increase in size, becoming larger vessels called veins.

Veins are thin-walled tubes that bring blood back to the heart. The returning blood inside the veins is under low pressure. Muscle contractions in the legs and other parts of the body help move this blood under low pressure back to the heart. Veins also have valves inside of them to prevent the possible backflow of blood.

The Blood

Human blood is liquid with solid materials floating within this medium. The adult human body contains about six liters of blood. The blood can be broken into two main parts for study: the liquid plasma and the cellular components.

The liquid plasma is composed mainly of water, but also contains materials such as hormones, antibodies, vitamins, salts, and food. The main function of the plasma is to transport most of the mentioned materials to proper locations within the body. Plasma is basically a clear liquid with a slightly yellow color. Plasma is not red like whole blood would appear because of the absence of the red blood cells. About 55% of your total blood volume is composed of plasma.

The cellular components of blood include the red blood cells, the white blood cells, and the platelets. The **red blood cells** are made in the bone marrow, and shortly after they are created they lose their nuclei. A red blood cell is often called a ***corpuscle*** because it lacks a nucleus. Red blood cells are also called ***erythrocytes.*** Red blood cells are red in color because they contain the chemical ***hemoglobin***, which is red in color because of the element iron present within its structure. Hemoglobin is a chemical that helps red blood cells hold much more oxygen than they could without this chemical. The red blood cells then distribute oxygen to all the cells in the body. The function of red blood cells is to transport oxygen and carbon dioxide in the body. Red blood cells live for about four months in the body, and then they are destroyed in the spleen and liver and replaced by the bone marrow. There are about 5 million red blood cells per milliliter of blood.

There are many different types of **white blood cells** in the body. White blood cells are often called ***leukocytes***. White blood cells act as an internal defense system against any invaders inside the human body. White blood cells are produced in the bone marrow, but also in the spleen and lymph nodes. There are about 5,000 white blood cells per milliliter of blood. Some white blood cells will leave the transport system and enter the surrounding tissue to attack pathogens that have invaded the body. The function of white blood cells is to protect the body against foreign substances. In Lesson 7-9, the role of the white blood cells and the immune system will be covered in more detail.

Platelets are also produced in the bone marrow. Platelets are formed when large cells from the bone marrow shatter into many small pieces. Platelets help the blood to clot in the event of an injury. When a blood vessel ruptures, platelets will stick to the jagged edges of these injuries. The platelets will then initiate the production of an enzyme called ***thrombin***. Thrombin produces a chemical that causes ***fibrinogen*** to be converted into the ***fibrin***. Fibrin, a sticky thread-substance, traps red blood cells and white blood cells and forms a seal over the ruptured blood vessel. Platelets live for about one week in the body before they are replaced.

Lymphatic System

The small blood vessels called the capillaries are responsible for most of the diffusion of materials into the surrounding tissue. The small size of these capillaries can, at times, also cause the loss of plasma into the body tissue. The **lymphatic system** is a series of tubes located throughout the body that helps collect this plasma fluid that has leaked out of the blood vessels into the surrounding body tissue. The name of this lost fluid when present in the lymphatic vessels is known as **lymph**. All these small lymph vessels eventually join to form larger lymph vessels, which empty the collected fluids into large veins that enter the heart near the neck.

The lymphatic system contains many oval-shaped structures called **lymph nodes**. These lymph nodes give rise to special white blood cells and antibodies. These nodes act as defense areas scattered throughout the body in such places as the neck, armpit, and groin. These lymph nodes can become swollen when the body is actively fighting a pathogen. The swelling of the lymph nodes in the neck, which is sometimes incorrectly called "swollen glands," is usually a sign of disease.

The lymph vessels are also the tubes by which fats enter the bloodstream from the small intestine. Protruding from the lining of the small intestine are the hairlike extensions called villi that contain the vessels called lacteals. These lacteals are part of the larger lymphatic system and absorb fat from the food.

Practice Section 7-2

1. Human red blood cells transport
   a) oxygen b) dissolved food c) hormones d) antibodies
2. The plasma is composed mainly of which chemical?
   a) alcohol c) sodium chloride
   b) water d) hormones
3. Which structures are responsible for the clotting of blood?
   a) white blood cells c) platelets
   b) red blood cells d) antibodies
4. Which chamber of the heart pumps oxygenated blood out to the parts of the body?
   a) right atrium b) left atrium c) right ventricle d) left ventricle
5. What is the name of the group of blood vessels that contain valves and bring blood to the heart?
   a) arteries b) veins c) capillaries d) lymph tubes
6. What is the name of the enzyme, produced from platelets, that begins the clotting process after injury?
   a) lymph b) thrombin c) amylase d) protease
7. In which blood vessels does the diffusion of most substances occur between cells and the surrounding tissue?
   a) arteries b) veins c) capillaries
8. What is the name of the chemical found in red blood cells that gives it its red color?
   a) hemoglobin c) lipase
   b) sodium chloride d) chlorophyll
9. Leukocytes is the biological name given to
   a) platelets b) red blood cells c) white blood cells d) lymph tubes
10. Pulmonary circulation of blood is the transport of blood between the heart and the
    a) liver b) kidneys c) lungs d) brain
11. What is the correct pathway the blood follows through the heart?
12. What is the difference between veins and arteries?
13. What is the purpose of the lymphatic system?

Lesson 7-3: Human Breathing

The life process of respiration was covered in Lesson 3-5 and involved the chemical processes required to obtain energy from the oxidation of food molecules. In this section we will look more closely at how the human body obtains the oxygen needed for the aerobic respiration process and how the body removes the waste products produced by this process.

The Human Respiratory System

The human body is covered in a protective layer of skin. This layer of skin keeps bacteria and other pathogens from invading the internal structures of the body and also prevents the loss of fluids. But this skin layer also makes it very difficult for all the internal cells of the body to receive the proper amount of oxygen needed for the process of respiration. The structures of the human respiratory system bring oxygen from the external environment to the internal cells.

Major Parts of the Human Respiratory System			
Nasal Cavity	**Bronchioles**	**Oral Cavity**	**Alveoli**
Trachea	**Epiglottis**	**Bronchi**	**Diaphragm**

••••••

Air from the environment enters the human body via the **nasal cavities** and the **oral cavity**. The nasal cavities inside the nose are lined with hairs and mucus, which traps dirt and bacteria that may enter with the incoming air. The air is also warmed to body temperature when it enters the nostrils. The mouth does not clean and prepare the external air as well as the nasal cavities, which means breathing in through your nose is usually better for the body.

The back area of the oral cavity above the esophagus is called the ***pharynx***. The ***tonsils***, which help protect the throat area from bacterial infection, are located in the pharynx. The **trachea**, or windpipe, and the esophagus meet at the end of the pharynx. The trachea is a tube that transports air from the oral cavity, during inhalation, to the lungs. A small flap of skin called the **epiglottis** automatically covers the opening to the trachea during swallowing, preventing food and liquids from accidentally going down the air pipe. Right below the epiglottis in the throat is the ***larynx***, or voice box, which is a collection of cartilage that produces sound when air rushing by causes this tissue to vibrate.

The trachea, as mentioned earlier, is a tube about 11 centimeters long that connects the oral cavity to the lungs. Surrounding the entire length of the trachea are rings of cartilage that prevent the windpipe from collapsing during the inhalation and exhalation of air. Cilia protruding from cells along the entire inner surface of the trachea sweep particles of dirt and other foreign substances back up into the pharynx, preventing them from entering the lungs. The windpipe branches into two tubes called **bronchi**, which extend to each of the lungs in the chest cavity. The bronchial tubes continue into the lungs and continue to decrease in diameter and continue to branch outward. These smaller tubes of the bronchi are called **bronchioles**. These many small bronchioles in the lung extend outward like branches of a tree, and end in a collection of very small ball-shaped structures called **alveoli**.

Human Respiratory System

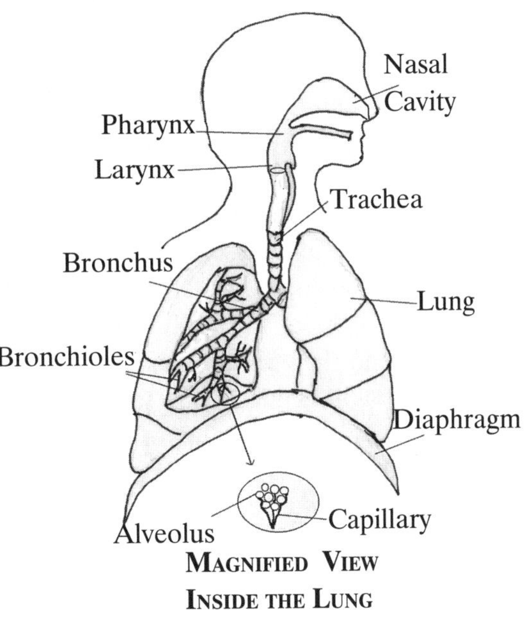

MAGNIFIED VIEW INSIDE THE LUNG

Alveoli are the locations where the gases from the environment are exchanged with the blood in the body. In the average human body there may be as many as 1 billion alveoli creating a huge surface area for the exchange of respiratory gases. Blood is brought close to these internal respiratory surfaces by the many capillaries that surround each alveolus. Diffusion of oxygen from the alveoli into the blood and carbon dioxide out of the blood into the alveoli is accomplished because the blood vessels and the structures of the alveoli are so thin in this area of the lungs.

Mechanical Process of Breathing

Each lung consists of the bronchi, bronchioles, and the many alveoli. The lungs do not have any muscles. Muscles outside of this structure control the movement of air into and out of the tubes of the lungs. One of the most important muscles involved in the movement of air into and out of the lungs is the **diaphragm**. The diaphragm is a sheet of skeletal muscle located below the lungs that aids in the act of breathing. When someone inhales, the diaphragm contracts and this flattens this muscle, moving it lower in the chest cavity. A slight vacuum is created inside the chest area, creating a space of lower pressure inside the chest compared to outside the body. Air floods into the lungs where there is lower pressure than the external pressure of the environment.

When someone exhales, the diaphragm relaxes and this brings this muscle back to its normal resting position. The space in the chest cavity becomes smaller during this action, causing the air pressure inside to increase. Now the air pressure is larger inside the chest than outside the body, and the air rushes out of the lungs into the environment.

The movement of oxygen in the blood

Once oxygen has entered the human lungs it must be brought to all the millions of cells found inside the body. The blood is the transport system for this oxygen. Human blood is red because it contains the chemical hemoglobin. Hemoglobin is an iron-containing compound that enables the red blood cells in the blood to hold onto more oxygen then the cells could without this chemical. In fact, red blood cells that contain hemoglobin can hold 60 times more oxygen then red blood cells that do not have hemoglobin. Where the capillaries meet the alveoli, through the process of diffusion, oxygen enters the red blood cells and binds with the hemoglobin. The red blood cells are transported by plasma in the blood vessels to all the different cells of the body. The oxygen leaves the hemoglobin and the red blood cells and then enters these oxygen-depleted cells. The chemical process of respiration has also filled these same oxygen-depleted cells full of carbon dioxide. When the red blood cells release oxygen, they pick up carbon dioxide. The carbon dioxide is brought to the alveoli and released into the lungs, and then during the next exhalation these gases are expelled into the environment.

Practice Section 7-3

1. The mechanical process of the human lungs bringing air into the body is called
   a) respiration
   b) breathing
   c) dehydration synthesis
   d) oxidation
2. Another name for the human windpipe is the
   a) epiglottis b) esophagus c) trachea d) larynx
3. The actual exchange of air into the blood and carbon dioxide into the lungs occurs in the
   a) trachea b) bronchi c) bronchioles d) alveoli
4. As blood passes through the lungs, gases are exchanged with blood vessels called
   a) arteries b) veins c) capillaries d) lymph tubes
5. When the diaphragm is bending into a downward curved position air
   a) enters the lungs b) exits the lungs c) there is no net movement of air
6. What is the function of hemoglobin in the blood?
   a) to help fight infection in the body
   b) to bind with food molecules for digestion
   c) to enable red blood cells to hold more oxygen
   d) to replace worn-out red blood cells

7. What muscle is responsible for breathing? How does this muscle work?
8. How is oxygen transported throughout the human body?
9. What is the difference between breathing and respiration?

Lesson 7-4: Human Excretory System

Metabolism of food and other chemicals in the body produces large amounts of waste products. The human excretory system consists of all the organs that aid the body in removing wastes produced from the processes of cell metabolism. It is important to stress, before we continue with this section, that excretion is the removal of metabolic waste. Some of the waste removed from the human body via the large intestines, rectum, and anus is not considered to be metabolic waste. Metabolic wastes are the unneeded chemicals produced from processes such as respiration, dehydration synthesis, and protein breakdown within the cells of the body. Wastes like carbon dioxide, water, ammonia, and salts are collected from the cells and are removed by way of the urinary system. The waste, which exits the anus of humans, is undigested food, or ***feces***, which means it has not entered the cells within the body. Removal of this type of waste is called ***egestion*** and is not considered to be part of the life process of excretion.

Metabolism of Proteins

Nitrogenous wastes produced from the chemical digestion of proteins are the most common and the most toxic wastes produced by the body. The liver filters old red blood cells from the body and is also responsible for the breakdown of amino acids. During the process of nutrition if too much protein is ingested and broken down into amino acids, it must be removed from the body because, unlike molecules of carbohydrates and lipids, it cannot be stored by the cells of the body. In the liver, an amino acid molecule is converted into ammonia and the molecule you see here.

Deamination of an Amino Acid

Amino group is removed

$$H_2N-C(H)(R)-C(=O)OH + H_2O \longrightarrow NH_3 + O=C(H)(R)-C(=O)OH$$

Ammonia is a highly toxic substance, so it is quickly converted into another chemical called ***urea***. Urea is a water-soluble material that is formed when ammonia is chemically combined with carbon dioxide in a series of chemical reactions. Urea is deposited into the blood stream from the liver and will eventually be excreted by the urinary system.

Structures of the Human Excretory System

Major Parts of the Human Excretory System

Kidneys **Bladder** **Ureters** **Urethra**

••••••

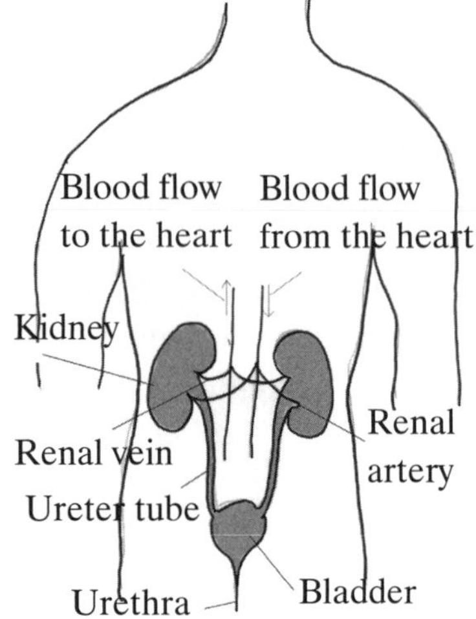

The **kidneys** are the filters of the human blood, removing the nitrogenous wastes produced from the liver, excess water, and salts. These wastes then exit the kidneys by way of the ureter tubes into the bladder. The bladder stores these wastes until they exit the body by way of the urethra.

Nephrons

Each kidney is actually a collection of millions of tiny filters called **nephrons**. Each nephron is a microscopic tube about 3 cm long that filters the waste products from the blood. Blood is brought into a collection of capillaries called the **glomerulus**, which, under high pressure, squeeze the water, food, salts, and urea from the blood into the surrounding structure called **Bowman's capsule**. All the materials deposited into Bowman's capsule move through the coiled tubes of the nephron. The food, some salts, and most of the water is reabsorbed back into the blood stream at a section of the coiled nephron tube called the **Loop of Henle**. This means that the only remaining substances in the nephron tubes after this point are nitrogenous wastes, some salt, and some water that collectively is called ***urine***. The blood leaving the nephron has been filtered of waste without the loss of much water or dissolved food molecules. Millions of microscopic droplets of waste from each nephron collect in the ureter tubes, leading out of the kidney and into the bladder.

Nephron

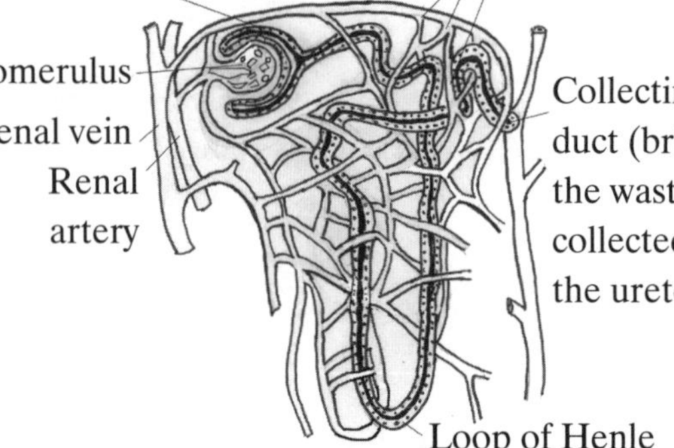

It is important to note that through the pores of the skin humans also release small amounts of urea, water, and salt.

Practice Section 7-4

1. The removal of waste from the body produced from cell metabolism is called
   a) egestion b) excretion c) synthesis d) deamination
2. Nitrogenous wastes are produced from the breakdown of
   a) carbohydrates b) proteins c) lipids d) simple sugars
3. What organ in the human body filters the blood of cellular waste?
   a) liver b) kidney c) large intestine d) pancreas
4. What is the name of the functional unit of the kidney?
   a) urethra b) nephron c) ureter d) dendrite
5. Which of the following chemicals are not found in human urine?
   a) water c) uric acid
   b) urea d) all are found in human urine
6. In what structure of the nephron is food reabsorbed back into the blood stream?
   a) glomerulus c) Loop of Henle
   b) Bowman's capsule d) bladder
7. What is the main function of the kidney?
8. Describe the structure of a nephron.
9. What is the difference between the process of egestion and excretion?

Lesson 7-5: Human Nervous System

Probably the most amazing and least understood of the human systems is the nervous system. The human nervous system in conjunction with the endocrine system helps regulate the many chemical and physical reactions occurring inside the body.

Nerve Cells

The human nervous system is composed of an elaborate network of cells called neurons. **Neurons** are the basic units of the nervous system. The figure on page 212 shows a diagram of a neuron.

These neurons, or nerve cells, are only visible with a microscope, but when collected in large numbers, they compose what is known as nerves, which are visible to the unaided human eye. The ***cyton***, or the body of the nerve cell, contains the cytoplasm and the other organelles needed to carry on the life processes. The hair-like extensions of the cyton are called the ***dendrites***. The dendrites receive stimuli

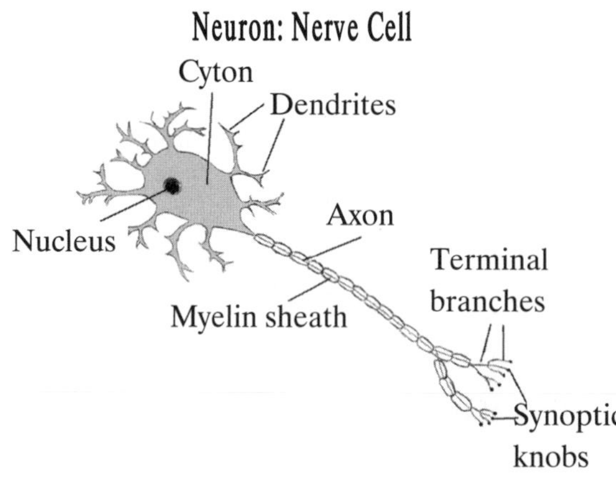

sent from another nerve or the outside environment. The **axon,** which is usually the longest part of the nerve cell, transmits the nerve impulse to the end of the nerve cell. **Myelin sheath** covers the axons of some nerve cells, and by doing so speeds the transmission of the nerve impulse. The axon ends in forked structures called **terminal branches**. Terminal branches send the nerve impulses to other nerves, organs, or glands within the body by the means of chemicals called neurotransmitters.

How Is a Nerve Impulse Transmitted?

The dendrites of the neurons are present in the eyes, ears, tongue, nose, and skin of humans. These dendrites receive various stimuli such as light energy, sound energy, chemical signals, and pressure from the external physical environment. These various stimuli trigger the dendrites of the neurons, which will in turn send a nerve impulse to the proper structure within the body. It is important to understand that the stimuli itself does not travel down the length of these nerves, but triggers the nerves to send an electro-chemical signal along this pathway.

To fully understand how a nerve is able to accomplish this task, let us first look at what a neuron looks like when it is not transmitting a signal. The cell membrane of the neuron pumps sodium ions outside of the cell and also pumps potassium ions into the cell. These ions would normally redistribute themselves back to their normal positions but the cell membrane does not allow this to occur. Only the potassium ions are able to slowly leak back outside of the cell, while the sodium ions after having been pumped to the outside of the cell membrane are forced to remain there. The larger number of positive ions on the outside of the cell membrane as compared to the inside of the cell membrane creates a ***polarized membrane***. A polarized membrane means that the outside of the cell membrane takes on a positive charge and the inside of the same membrane takes on a negative charge. When a nerve impulse is sent down the resting neuron the following actions occur. A chemical signal is sent to the dendrites of the neuron causing it to become ***depolarized***. Depolarization is when a sudden change in the permeability of the cell membrane allows the sodium ions to rush into the cell. This causes the charges on the membrane to reverse, with the inside of the membrane becoming positive while the outside becomes negative. Shortly after this occurs, the sodium-potassium pump restores the ions to the resting neuron locations.

The area of depolarization lasts for a very short period of time and causes the area next to depolarized cell membrane, in the direction of the terminal branches, to undergo the same chemical process. Nerve impulses always move along the neuron in the direction toward the terminal branches. The initial stimulation of the dendrite section of the neuron must be large enough to start this electro-chemical event. The needed amount of stimulation is called the ***threshold potential***.

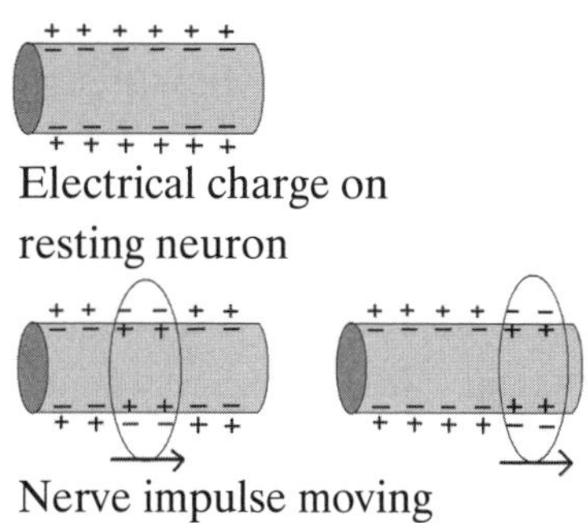

Electrical charge on resting neuron

Nerve impulse moving down the nerve cell

Strength of the Impulse

Very important to the function of neurons is that the strength of the stimuli sent to the various body structures from the environment is distinguishable. This may sound like a pretty basic assumption, but the electro-chemical nerve impulses that are sent down the neuron always have the same strength. How, then, is this difference in impulse strength accomplished? The answer to this question lies in the rate of nerve impulse transmission along the neuron. For example, if someone taps you on the arm, a low number of nerve impulses per second are sent from the dendrites in this area to the central nervous system. The number of impulses sent per second would be much higher if you were punched in that same area.

The number of neurons, which are involved in the reception and transmission of this impulse, also control the strength of an impulse. Using the same example mentioned previously, when your arm is slightly tapped, fewer neurons are usually stimulated; while when you are hit harder, many more neurons are stimulated and they send an impulse to the central nervous system.

Impulse Transmission From Nerve to Nerve

The human body is filled with an elaborate array of neurons that bring impulses to all parts of the body. A very interesting fact about this arrangement is that no neuron that is transmitting an impulse to another neuron actually touches. Between the dendrite of one neuron and the terminal branches of another there is actually a small space called a **synapse**. The depolarized action potential created on the neuron cannot jump across this synapse. So when the impulse reaches the end of the terminal branches of one neuron, the ends of these branches secrete chemicals known as **neurotransmitters**. These neurotransmitters diffuse across this space and bind to receptor sites on the dendrites of the next neuron. When the receptor sites of the dendrites are triggered by these neurotransmitters, the action potential continues at that point on the cell that continues the impulse. Why would these spaces called synapses be an advantage for proper nerve functioning within the body? This space provides a way for the nerve impulse to be regulated inside the body. Many neurons share common synapse spaces within the body. This means that usually many impulses are sent to a common synapse at the same time, which also releases many neurotransmitters into this space at the same time. There are different kinds of neurotransmitters secreted into these spaces, some that cause the nerve impulse to continue, and others that cause the nerve impulse to stop. The final verdict of where and when these impulses travel is determined by the net results of these many neurotransmitters secreted into these synapses. This enables the many complicated reactions needed by higher organisms to occur. A good example of how two nerves leading to the same organ can have different affects on this organ is illustrated below.

The Terminal Branch of a Nerve Leading to the Heart Muscle Can Release Each of the Following Chemicals If It

Releases the neurotransmitter	Releases the neurotransmitter
Acetylcholine	**Noradrenaline**
The heart beats slower	The heart beats faster

Table 7.1

Reflex Arc

In all vertebrates there are nerves that run to and from the central nervous system forming what is called a **reflex arc**. The human body is covered in receptors that can be bare dendrites in the skin or nerve endings in one of the many sense organs. When a stimulus triggers one of these bare nerve endings, an impulse is transmitted by way of **sensory neurons** to the central nervous system. Another type of nerve cell called an **interneuron**, located in the central nervous system, then picks up the impulse from the sensory neuron. Then the interneuron transmits the impulse to a motor neuron that will bring this impulse to an effector (organ or gland) that will complete the reflex arc. For example, a baseball pitcher throws the ball over the plate to a batter and the batter hits a line drive toward the head of the pitcher. The nerve endings in the eye (receptor) will collect the light images of the ball shooting toward the head of the pitcher. A nerve impulse will be sent along sensory neurons from the eye to the spinal cord of the central nervous system. In the spinal cord, the interneurons will transmit this impulse from the sensory neurons

Human Reflex Arc

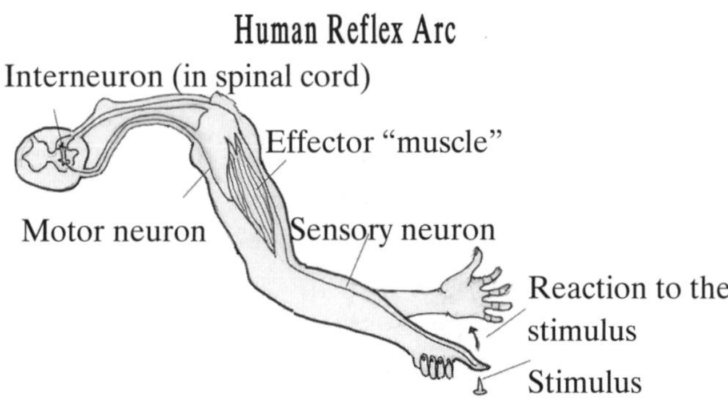

to the proper motor neurons in the body. These motor neurons will send signals to the muscles (effectors) in the head, neck, and back, causing the pitcher to duck just in time to miss being hit by the incoming baseball. Remember that there are spaces between all the nerves involved in the reflex arc synapses that keep the impulse moving through the release of neurotransmitters.

The nervous system of humans is divided into two parts: the **central nervous system** and the **peripheral nervous system**.

The Central Nervous System

In humans the central nervous system consists of the brain and the spinal cord. The spinal cord is a hollow tube made of many nerves that carry impulses to and from the brain from the rest of the body. The 26 vertebrae of the backbone protect the spinal cord. The spinal cord not only helps transport nerve impulses around the body, but also controls the reflex actions of coughing, sneezing, blinking, and movements such as pulling your finger back quickly when it hits something hot or sharp.

The highly developed brain is probably the organ that most separates humans from all other forms of life on this planet. A thick layer of bone called the cranium, or skull, covers the brain. Three soft tissue layers called ***meninges*** sit between the skull and the brain. The meninges cushion the brain against jarring movements against the skull.

Parts of the Human Brain

The brain has three major sections or divisions.

The **cerebrum** is the largest portion of the brain that is actually separated into two halves called ***cerebral hemispheres***. The cerebrum controls such abilities as thought, creativity, and reasoning, which is why this part of the brain is so much more highly developed in humans than in most other animals. There are sections of the cerebrum that interpret nerve signals sent from the various sense organs of the body such as the eyes, tongue, and ears, while there are also areas of the cerebrum that aid in the voluntary movement of muscles in the body. The outside of the cerebrum is called ***cortex*** tissue and contains what is sometimes called the ***gray matter*** of the brain. Gray matter consists mainly of the cyton portions of the nerve cells. The inner section of the cerebrum contains white matter and consists mainly of the axon portions of the nerve cells.

Structures of the Human Brain

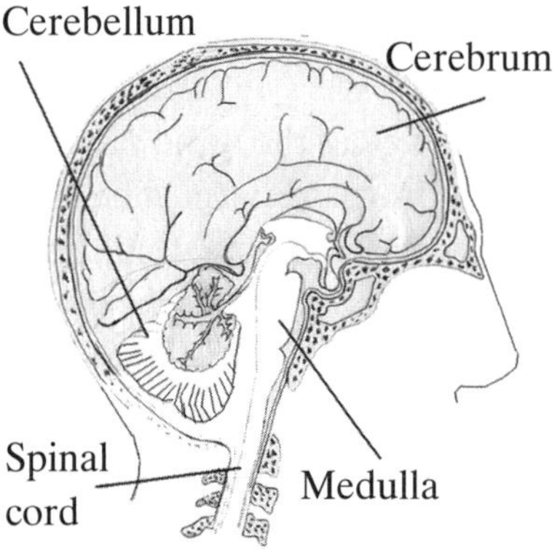

The **cerebellum** is much smaller in volume than the cerebrum and is located in the lower back section of the cranium. This section of the brain controls the coordination of voluntary muscles and the ability to keep the body balanced. For example, the ability for a baseball player to swing a wooden bat and hit a 90-mph fastball involves the coordination of many muscles. It is interesting to note that the cerebellum of many birds is highly developed, which gives them the ability to balance on tree branches almost effortlessly.

The **medulla**, which is sometimes called the medulla oblongata, is located underneath the cerebrum at the top of the spinal cord. This part of the human brain controls most of the basic processes needed to sustain life—breathing, the beating of the heart, and the movement of food through the digestive tubes.

Three Major Sections of the Human Brain		
Cerebrum	**Cerebellum**	**Medulla**

●●●●●●

The Peripheral Nervous System

The **peripheral nervous system** is divided into two parts: the ***somatic nervous system*** and the ***autonomic nervous system***. All the nerves that are found within the body but are outside of the brain or spinal cord are considered part of the peripheral nervous system.

The somatic nervous system transmits all the impulses between the voluntary muscles of the body and the spinal cord and brain. There are 12 pairs of nerves extending from the brain to various structures in the face, such as the nose, ears, and eyes. There are 31 pairs of nerves extending from the spinal cord to various structures that need voluntary control within the body. The somatic nervous system contains both sensory and motor neurons.

The autonomic nervous system transmits nerve signals to many of the internal organs of the human body that need to be controlled without thought. Many of the actions of the circulatory system, the respiratory system, and the digestive system are controlled by the autonomic nervous system. The autonomic nervous system consists of the ***sympathetic*** and the ***parasympathetic systems***. These systems both send nerves to the same parts of the body, but each of these systems has opposite effects on these parts. The sympathetic nerves increase heart rate and releases hormones, such as adrena, for possible actions the body may have to undergo. The parasympathetic nerves calm the body by stopping the release of certain hormones and decreasing heart rate and breathing. These two nerve systems can have opposite effects on the same organ by releasing different neurotransmitters into the synapses leading to the organ in question. The nerves of the autonomic nervous system are only motor neurons and form connections between the spinal cord and the organs of the body they control.

The Effects of Drugs on the Nervous System

Alcohol

Alcohol is fermented types of fruit juice, or grains. Alcohol that is created in this manner is called ***ethanol***, or ethyl alcohol.

Alcohol is a ***depressant*** to the central nervous system. This means that alcohol slows down all the functioning rates of the central nervous system. Alcohol also slows the secretion of certain neurotransmitters of the nervous system. Effects of low doses of alcohol to the human body include: slowing of reflexes, impairment to concentration and reaction time, lowering of inhibitions, reduction of stress and tension, and reduction of coordination. In medium to high doses, alcohol produces the following effects: slurred speech, drowsiness, vomiting, breathing difficulties, coma, and even death.

Marijuana

Marijuana comes from a plant with the scientific name ***Cannabis sativa***. The chemical in this plant species that produces the effects on the nervous system is called delta-9 tetrahydrocannabinol or THC. Marijuana is usually smoked like a cigarette. Marijuana is a depressant to the nervous system. The chemicals in marijuana act on receptors in the brain that are responsible for concentration, memory, perception, and movement. Symptoms produced in the human body by the chemicals in this drug include reduced coordination, reduced blood pressure, disorientation, hallucinations, and impaired memory. When smoked, this drug also has many harmful effects on the lungs, similar to cigarettes.

Cocaine

Cocaine is a drug produced from the plant with the scientific name ***Erythroxylon*** *coca*. Cocaine is obtained from the leaves of that plant. Cocaine is a central nervous system stimulant. It can be taken by chewing on coca leaves, and it can be smoked, inhaled, or injected. Cocaine releases a neurotransmitter called ***dopamine*** in the brain. Dopamine is usually secreted to tell the body a basic need has been met. Dopamine produces intense feelings of satisfaction and also greatly increases heart rate and blood pressure.

The drug can produce a psychological dependency, because shortly after the effects of the drug have worn off, the user may feel sad or depressed because of the depletion of dopamine in the system. This causes the person to want to use more cocaine to recapture the "high."

Practice Section 7-5

1. What part of the neuron receives the impulses sent from another nerve cell?
   a) cyton
   b) myelin sheath
   c) terminal branches
   d) dendrites

2. What chemicals are responsible for creating the charge present in the formation of an impulse in a neuron?
   a) oxygen and carbon
   b) potassium and sodium
   c) nitrogen and sodium
   d) potassium and oxygen

3. Which of the following conditions determines how the body interprets the strength of a particular stimulus from the environment?
   a) the number of impulses sent per second down the neuron
   b) the number of neurons that are stimulated from the stimuli
   c) both A and B
   d) none of the following

4. How does the nerve impulse travel from neuron to neuron?
   a) an electrical charge jumps the gap between neurons
   b) all neurons are connected so the impulse travels smoothly along the axon
   c) neurotransmitters are released from one neuron to the other
   d) an electrical charge travels from the dendrite of one nerve to the terminal branch of the other

5. What type of nerve sends the impulse from the environment to the central nervous system?
   a) sensory neuron
   b) motor neuron
   c) interneuron
   d) neurotransmitter

6. What section of the central nervous system is responsible for the control and coordination of voluntary muscles and balance?
   a) cerebrum b) cerebellum c) medulla d) spinal cord

7. Define the term *reflex*. Describe the operation of a reflex arc.

8. What are the two systems of the autonomic nervous system? What are their functions?

9. What are the gray matter and white matter of the cerebrum made of?

Lesson 7-6: Human Endocrine System

The human endocrine system controls the flow of **hormones** throughout the body. Hormones are chemicals secreted by various glands that travel to their target location by way of the human transport medium: the blood. Control of the various systems within the body is accomplished by both the endocrine and nervous system. In many cases, these two systems function together to achieve the desired action. The nervous system controls the body with quick, but short-lived actions while the endocrine system is much slower to react, but whose effects usually last much longer. The coordination and control of various body functions by both the endocrine and nervous systems is called homeostasis.

The Major Parts of the Endocrine System

Pituitary Gland	**Adrenal Gland**
Thyroid Gland	**Islets of Langerhans**
Parathyroid Gland	**Testes**
Thymus Gland	**Ovaries**

••••••

The Nervous System and Endocrine System Connection

The ***hypothalamus*** is a very small structure located at the bottom of the brain and is considered part of the central nervous system. This part of the brain is connected to the pituitary gland, also located in the skull, by a series of nerves. Nerve signals are sent by the hypothalamus to the pituitary gland, causing the release of chemicals. These chemicals released by the pituitary are then sent to many of the other glands in the body causing them to also react. This is why the pituitary gland is sometimes called the "master gland," because it can affect so many other glands within the body. The pituitary gland is composed of two parts: the anterior section and the posterior section. Each section of this gland secretes its own hormones. Table 7.2 shows the hormones released by the two sections of the pituitary gland, and the effects of these hormones on the body.

Anterior Pituitary Gland

Hormone	Action
(GH) Growth hormone	Stimulates the rate of growth during childhood.
(LH) Luteinizing hormone	Starts the ovulation process in females and helps form the *corpus luteum*.
(FSH) Follicle-stimulating hormone	Helps to mature the eggs in females and the sperm in males. This hormone also causes both the ovaries and testes to release sex hormones.
Prolactin	Stimulates milk production in the female mammary glands after pregnancy.
(TSH) Thyroid-stimulating hormone	Stimulates the thyroid gland to release its hormone, thyroxine.
(ACTH) Adrenocorticotrophic hormone	Stimulates the adrenal cortex to release its hormone, cortisol.

Table 7.2

The hormones defined in the follwing chart are actually produced in the hypothalamus of the brain, but are stored and released by the anterior pituitary gland.

Posterior Pituitary Gland

Hormone	Action
(ADH) Antidiuretic hormone	Stimulates the kidney to act in such a way as to reduce the amount of water that is lost in the urine.
Oxytocin	Stimulates uterine contractions during childbirth.

Table 7.3

After they are released by a specific gland, hormones travel in the bloodstream to a particular "target tissue." The hormones attach to specific sites in or on the cells of the target tissue called **receptor sites**. Any tissue that the hormone cannot bind to will not be affected; this is why after being released by a gland certain hormones only have effects on certain parts of the body.

Endocrine System Self-Regulates

The hormones secreted by the endocrine system are very powerful, even in small doses. This means that how they are used and in what quantity must be regulated very closely. The body achieves this close monitoring of its hormones by a ***feedback mechanism***. A feedback mechanism is when a gland in the body secretes a chemical that causes a change in another part of the body. This part of the body then sends a signal back to the gland to tell it to secrete more or less hormone.

General Feedback Loop

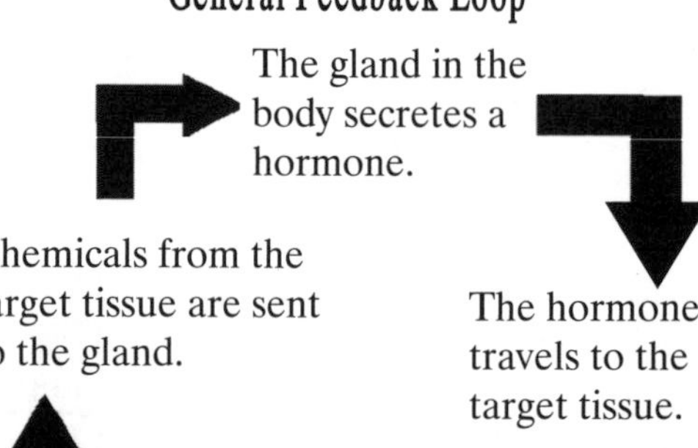
The gland in the body secretes a hormone.
Chemicals from the target tissue are sent to the gland.
The hormone travels to the target tissue.

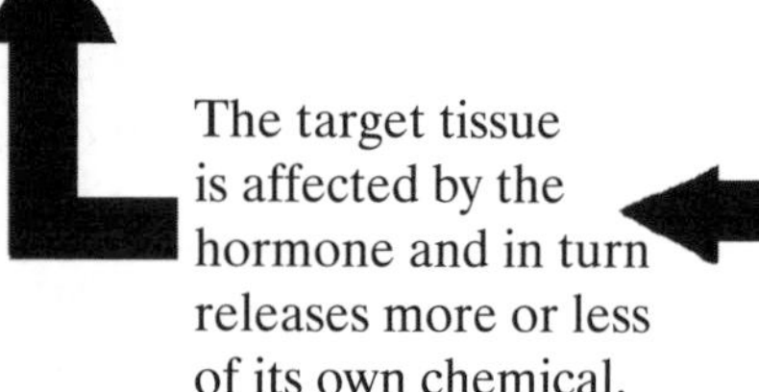
The target tissue is affected by the hormone and in turn releases more or less of its own chemical.

There are two types of feedback mechanisms; one is called ***positive*** and the other is called ***negative***. A positive feedback cycle is when a gland senses some chemical in the body and secretes a hormone that causes the target tissue to secrete *more* of the initial chemical released. Positive feedback cycles in the human body are rare. The an example of a positive feedback cycle in the body is shown on page 221.

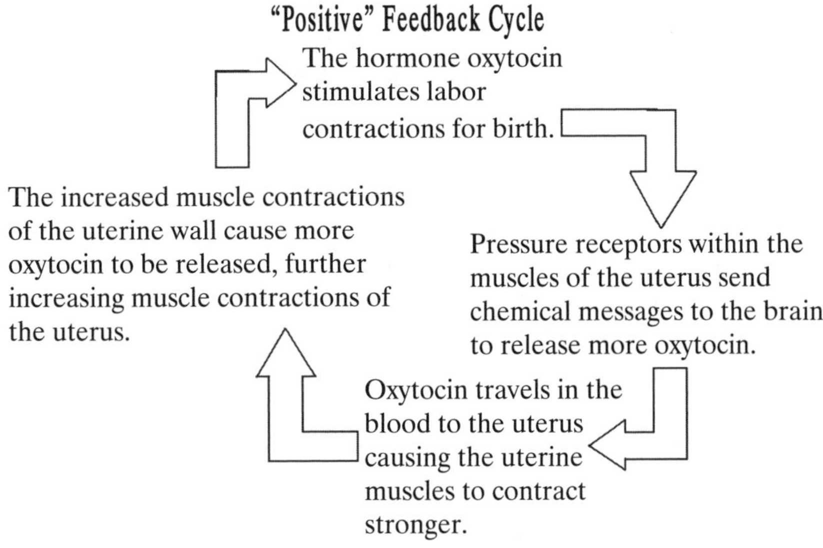

"Positive" Feedback Cycle
The hormone oxytocin stimulates labor contractions for birth.
Pressure receptors within the muscles of the uterus send chemical messages to the brain to release more oxytocin.
Oxytocin travels in the blood to the uterus causing the uterine muscles to contract stronger.
The increased muscle contractions of the uterine wall cause more oxytocin to be released, further increasing muscle contractions of the uterus.

A negative feedback cycle is when a gland senses some action in the body and secretes a hormone that causes the target tissue to secrete *less* of the chemical causing that action. An example of a negative feedback cycle in humans can be viewed in the following chart.

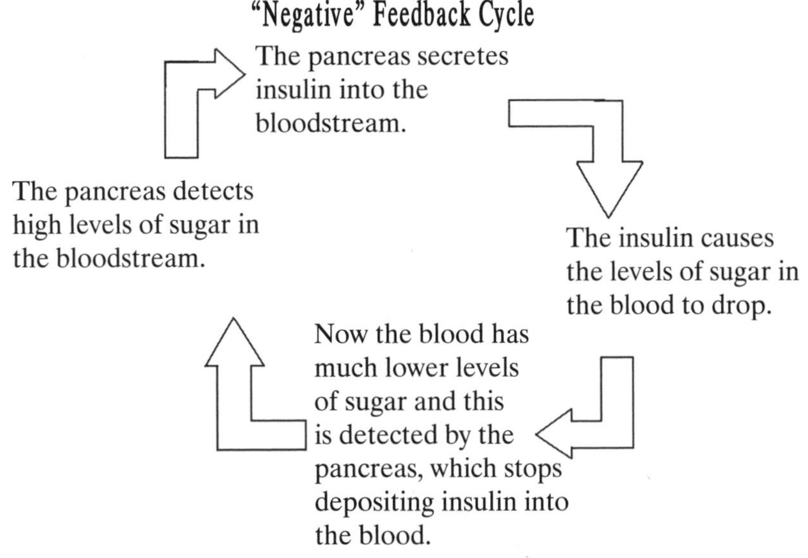

"Negative" Feedback Cycle
The pancreas secretes insulin into the bloodstream.
The insulin causes the levels of sugar in the blood to drop.
Now the blood has much lower levels of sugar and this is detected by the pancreas, which stops depositing insulin into the blood.
The pancreas detects high levels of sugar in the bloodstream.

Thyroid Gland

The thyroid gland is located in the neck curling around the trachea. This gland produces the hormone **thyroxin**. Thyroxin increases the rate of cell metabolism. Iodine is needed by this gland to produce the hormone thyroxin. **Hypothyroidism** is an undersecretion of thyroxin that lowers the rate of metabolism. The general physical symptoms of someone with this condition are sluggishness and being less mentally alert. In **hyperthroidism** there is an oversecretion of the hormone thyroxin that, in turn, increases the person's metabolic cell rate. The physical symptoms of someone with this condition are restlessness, high body temperature, and weight loss.

Parathyroid Gland

A group of four small glands called the parathyroid glands are attached to the thyroid gland. Parathyroid glands secrete the hormone **parathormone**, which regulates the amount of calcium in the blood and how much of this calcium enters the bones. Correct calcium levels in the body are important for blood clotting and the actions of muscles and nerve cells.

Islets of Langerhans

Special cells called the "Islets of Langerhans," which are named after the German physician Paul Langerhans, who examined these "island like cells" inside the pancreas, secrete the hormones **insulin** and the **glucagons**. Insulin lowers the amount of sugar in the blood by either causing the sugar to be absorbed by the cells in the body or by converting the sugar to glycogen and storing it in the liver. Glucagon increases the amount of sugar in the blood by releasing the stored sugar from the liver.

Diabetes mellitus is a condition where the amount of insulin produced is insufficient, causing large amounts of sugar to remain in the blood. This sugar is then excreted from the blood into the kidney as waste. The loss of large amounts of sugar from the body and the accompanying amounts of water can cause many organs to fail.

Adrenal Glands

There are two adrenal glands; one connected to the top of each kidney. Each adrenal gland has two parts: the outer ***adrenal cortex*** and the inner ***adrenal medulla***.

1. The inner adrenal medulla produces the hormones **adrenaline** and **noradrenaline**. The release of these hormones is stimulated by the nervous system. The hormone adrenaline causes an increase in heart rate and breathing, diverts more blood to the brain and less to the digestive organs, and also causes the liver to release more stored sugar into the blood. All these actions quickly prepare the body to respond to stimuli from the environment.
2. The outer adrenal cortex secretes steroid hormones, such as **cortisol**. Cortisol raises the level of sugar in the blood by converting proteins and fat into these sugars. Another hormone secreted by the adrenal cortex, called **mineralocorticoids**, affects the kidney in such a way as to help maintain blood pressure and water balance.

Gonads

The **gonads** are the sex organs of males and females that produce the gametes. In females, the gonads are called ovaries; in males, they are called testes. In males, the testes, which are responsible for the secondary sex characteristics, such as facial hair, a deeper voice, and broad shoulders, produce the sex hormone testosterone. This hormone also aids in the production of sperm.

In females, the ovaries produce the sex hormones estrogen and progesterone. Estrogen is the hormone responsible for such secondary sex characteristics as the enlargement of the breasts, a narrowing of the waist and enlarging of the hips, and the production of hair under the arms and around the vagina. The hormones estrogen and progesterone help in the build-up of the uterine lining for pregnancy.

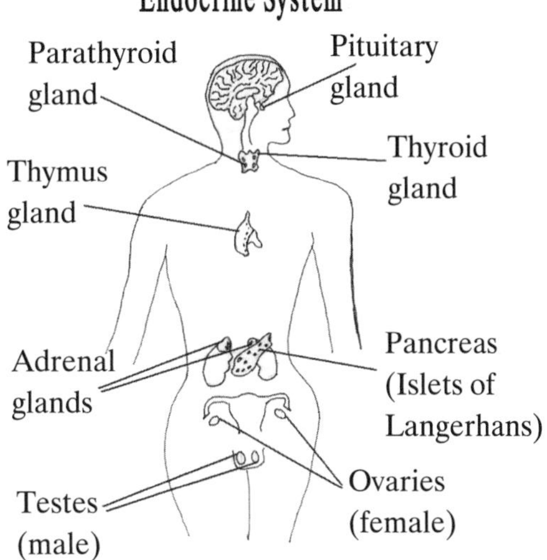

Practice Section 7-6

1. Which of the following glands is not part of the endocrine system?
   a) pituitary gland b) adrenal gland c) salivary gland d) thymus gland
2. Homeostasis in humans is regulated by
   a) the nervous system
   b) the endocrine system
   c) A and B
   d) none of these
3. The hormone insulin is produced in which body structure?
   a) pancreas b) liver c) stomach d) brain
4. When the human body is experiencing a stressful situation, which of the following hormones is secreted?
   a) thyroxin b) adrenaline c) oxytocin d) TSH
5. How are hormones transported throughout the body?
   a) through the bloodstream
   b) through the lymphatic ducts
   c) through special tubes connected to each gland
   d) through the bone structure

6. When a gland senses some action in the body and secretes a hormone that causes a target tissue to secret less of the chemical causing this action, this condition is called a(n)

   a) positive feedback cycle
   b) negative feedback cycle
   c) cyclosis
   d) aneurysm

7. What is a hormone?

8. What hormone is secreted by the thyroid gland, and what are its functions?

9. Where is the pituitary gland located in the body? List the major hormones secreted by the pituitary gland.

10. Name the hormones secreted by the ovaries and testes. What are the effects of these hormones?

Lesson 7-7: Human Senses

Interaction of the human body with the surrounding environment is critical to survival. The human sense organs are responsible for this job. The lights, sounds, smells, tastes, and touches from the world around us are received by the sense organs of the eyes, ears, nose, tongue, and skin. Sensory receptors within each of these sense organs allow the human nervous system to collect stimuli from the environment.

The Five Human Sense Organs				
Ears	**Eyes**	**Nasal Cavity**	**Nerves in the Skin**	**Taste Buds**

●●●●●●

Eyes and Sight

The sense organs that help collect light from the environment are the eyes. The structure of the eye is made of many different parts that help to collect, focus, and analyze light energy.

The Human Eye

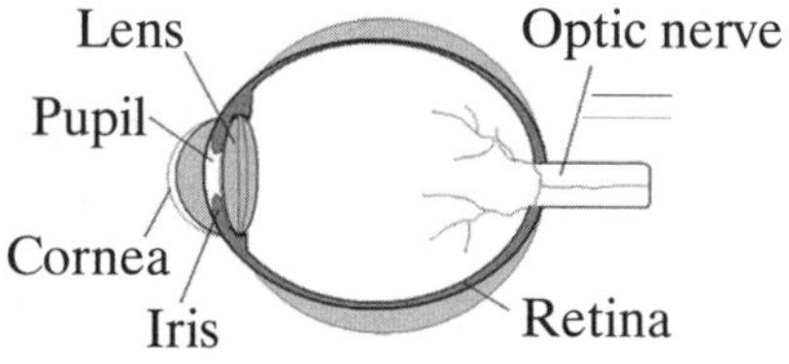

Lens
Optic nerve
Pupil
Cornea
Iris
Retina

The **cornea** focuses the light that enters the eye from the environment. The **iris** is the part of the eye that controls the opening to the inner eye. This opening in the iris is called the **pupil**. The pigments in the iris give humans their particular eye color. The iris can adjust the size that the pupil opens to let in more or less light. The **lens** in the inner eye helps the eye adjust the light coming off near or far-away objects. Many vision problems have to do with improper functioning of the lenses of the eyes. Light is projected from the lens of the eye to a structure in the inner eye called the **retina**. There are **photoreceptors** found in the retina that convert light energy into nerve impulses that will be sent to the brain by way of the **optic**

nerve. Two types of photoreceptors, cones and rods, control the collection of this light energy in the retina. Rods are very good collectors of light energy, even in low levels. Cones also collect light energy, but are able to distinguish the different colors of light. It is important to remember that the eyes collect, focus, and send the light energy to the brain. The brain is what truly interprets and gives meaning to the light energy. For instance, the many colors that are witnessed in the environment are just different wavelengths (or energy amounts) of visible light. It is the brain that associates color with them. So in some ways, the notion of color in the environment is just an illusion created by our brains.

Ears and Hearing

The human ears collect the stimulus of sound in the process of hearing. Sound is the movement of air molecules detected by a particular listening device. The human ears are a structure very well constructed to detect the movement of sound waves in the air.

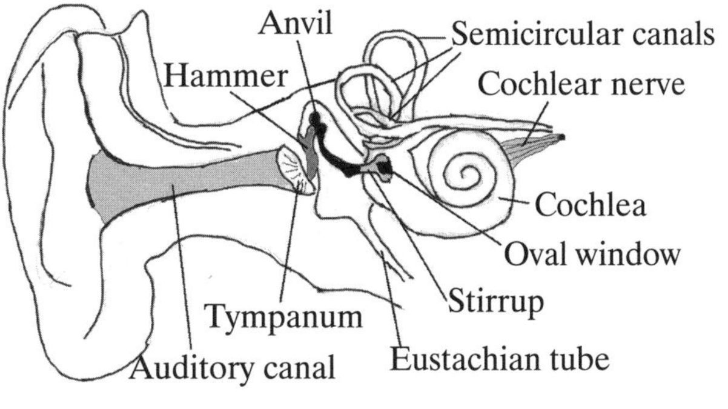

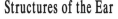

Waves of air molecules first enter the **auditory canal**. At the end of the auditory canal there is a thin membrane called the eardrum, or **tympanum**. The movement of air molecules causes the eardrum to vibrate. These vibrations are transferred to three tiny bones connected in order from the eardrum. These three bones are called the **hammer**, **anvil**, and **stirrup**. The vibrations move along these bones until they are transferred from the stirrup to the **oval window** of the **cochlea**. The vibrations from the stirrup cause waves within the fluids found inside of the cochlea. These fluid waves trigger very small fine hairs inside the cochlea. The stimulation of these hairs in the cochlea produces nerve impulses that are sent to the brain for interpretation.

It is also important to point out at this time that the ear helps the human body maintain balance. Above the cochlea are three structures called **semicircular canals**. These canals contain fluid and are lined with hairs. Depending on the position of the body, especially the head, the fluid in these structures will be found in different parts of the canal. The hairs lining the inside of these canals will be triggered depending on where the fluid is at any particular moment. These hairs then send nerve impulses to the brain, and the brain interprets the position of the body.

At the base of the middle ear is a tube that connects the middle ear with the back of the throat. This tube maintains atmospheric pressure on both sides of the eardrum. If air pressure on the outside of the eardrum changes, the action of swallowing or opening the mouth either releases a buildup of air in the middle ear or sends air into the middle ear to control air pressure balance. Sometimes you can sense this occurring when you have the sensation of your middle ear "popping." People who are at high elevations, such as aboard a plane or driving to the top of a mountain can experience this sensation. This "popping" is the middle ear adjusting to the changing air pressure.

Smell and Taste

You may be wondering why we cover the topics of smell and taste together in this section. It is because both of these sense organs are designed to collect chemicals from the environment. Chemical molecules that enter the nose and mouth are collected by receptors in these areas. The receptors in the nose for smell are very small, hairlike nerve endings. When chemicals contact these nerves, impulses are sent to the **olfactory nerve** in the brain. Unique chemical molecules will trigger certain nerve endings, while others will not. The brain interprets the nerve impulses sent by the various nerve endings as a particular odor.

The sense of smell is very closely linked to the sense of taste. When chemical molecules dissolve in the saliva of our mouths, they can trigger **taste buds** found on the lining of the four taste classes are interpreted more or less strongly on certain areas of the tongue. All the foods that we taste are combinations of these four.

Structures of the Human Tongue

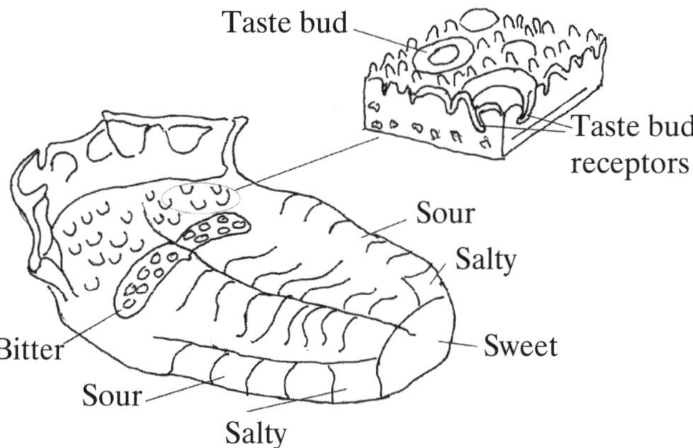

Chemicals that enter the mouth are also interpreted by receptors in the nose. Much of what we taste really is what we smell. Recall times when you had a bad cold and your nose was stuffy; it was very difficult to taste the food you ate.

Sense of Touch

Sense of touch is achieved by the millions of nerve endings located throughout your body in your skin. These nerve endings respond to the stimuli of pressure, pain, and temperature. Although the entire body is covered in nerve endings, areas such as the fingers, toes, and face have more nerves that are sensitive to softer stimuli.

The actual strength of all impulses sent from the skin to the central nervous system is the same. The strength of a nerve impulse from pressure to the skin depends on two things: the number of nerves endings in the skin actually stimulated by touch and the frequency of impulses that these nerves send to the central nervous system. This means that the strongest pain or sense of pressure to the skin causes the highest frequency and greatest number of impulses to be sent along the neuron.

Practice Section 7-7

1. Which of the human sense organs collects the stimulus of sound waves?
   a) eyes b) ears c) nose d) tongue
2. What nerve in the brain receives impulses from stimuli picked up by receptors in the nose?
   a) taste buds c) cochlea
   b) olfactory nerve d) auditory canal
3. If your nose is stuffy, what other sense organ will probably be affected?
   a) sense of sight c) sense of taste
   b) sense of touch d) sense of hearing
4. What part of the human ear aids in maintaining balance?
   a) cochlea c) semicircular canals
   b) tympanum d) anvil
5. What part of the body does not contain "touch" nerve sensors?
   a) fingers b) toes c) face d) brain
6. What part of the human eye distinguishes between the different colors of light?
   a) rods b) cones c) pupil d) lens
7. What structure inside the ear do humans use to hear? In what part of the inner ear is this structure located?
8. The eye is made of different structures that help you collect, focus and then analyze light energy? What structure does each one of these important steps in the process of sight?
9. What determines the actual strength of a stimulus perceived from the environment?

Lesson 7-8: Human Locomotion

In humans, movement is accomplished by the working of many different living tissues within the body. Bone, cartilage, muscles, ligaments, and tendons are all used for locomotion. All vertebrates have a hard, internal skeleton called an **endoskeleton**. The human skeleton is composed of cartilage and many different kinds of bones. The skeleton supports and maintains the shape of the animal, protects vital organs, and provides an area of support to which muscles can attach for movement.

Bones

Most of the human skeleton is bone. Bone is a living tissue. This means that bone is constantly growing. Many people think that bone is a nonliving substance, similar to a rock, because their experiences have been with bones from dead organisms.

The bones of humans are made of branching cells embedded in a matrix.

This matrix is approximately
30% protein and 70% mostly inorganic calcium phosphate.

• • • • • •

Bone comes in two forms: **compact bone** and **spongy bone**. Compact bone is more dense and solid than spongy bone and provides the support for the soft tissue of the body. The spongy bone is less dense than compact bone and is filled with soft tissues called ***red marrow***, which makes red blood cells, and ***yellow marrow***, which helps the body store fat.

Human Bone

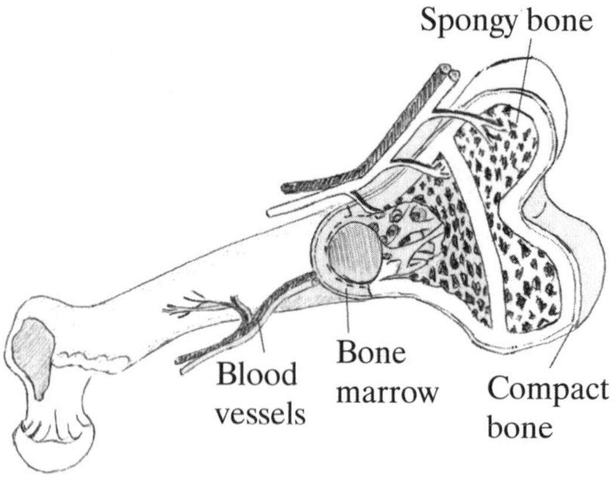

Human Skeleton

The human skeleton consists of 206 bones. The diagram on page 229 shows some of the different bones of the human body.

Human Skeleton

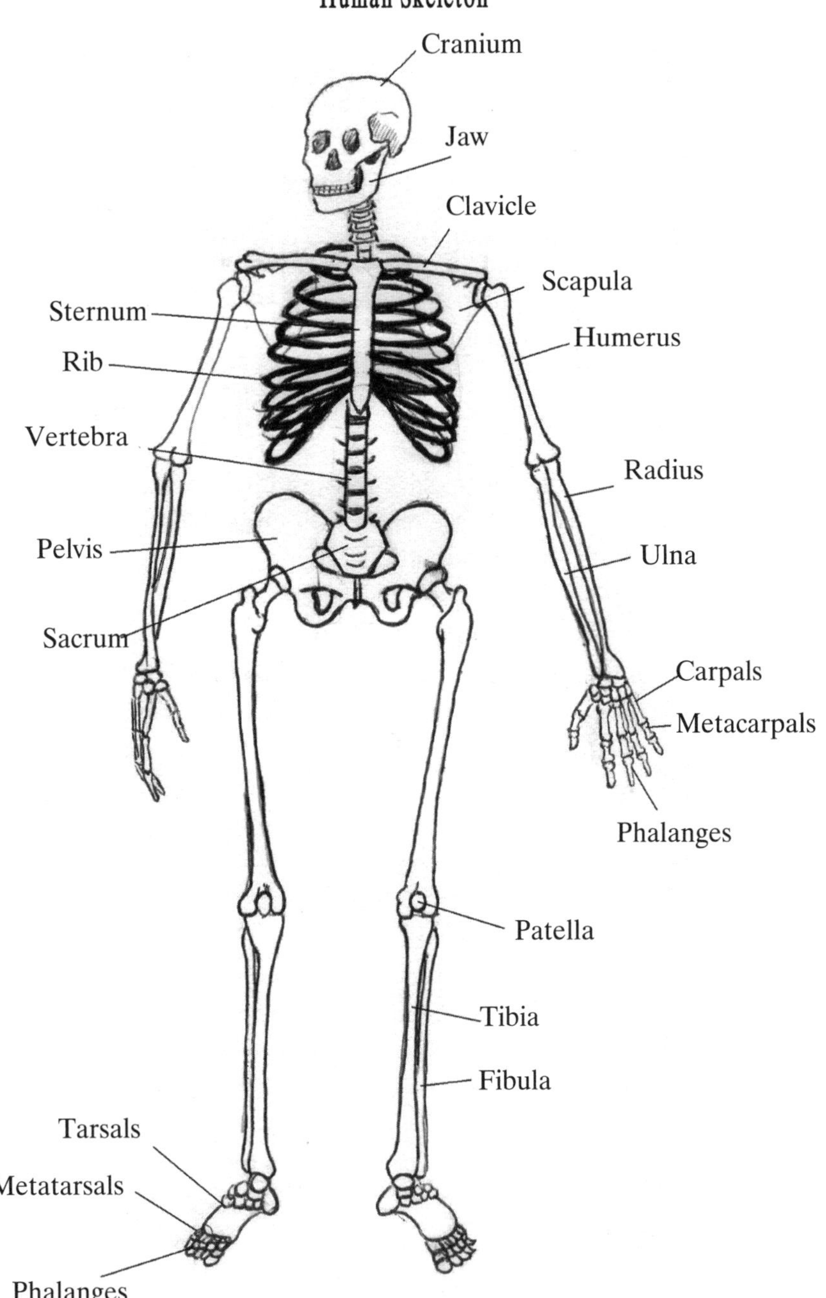

The many bones of the human skeleton are connected at points called **joints**. Most joints allow the bones to move at this connection point. Some joints are immovable.

The Joints of the Human Skeleton

Type of Joint	Description	Example
Fixed joints	No movement is allowed by these types of joints	Bones of the skull
Slightly movable joints	Bones meeting at these joints have some ability to move	The spaces between the vertebrate of the backbone
Freely movable joints	Bones meeting at these joints have the possibility of considerable movement	The joints between the arm bones

Table 7.4

Cartilage

The second type of tissue that makes up the human skeleton is a substance called **cartilage**. Cartilage is made of circular cells embedded in a rubbery matrix that has supporting fibers. In human adults, you will find cartilage in the ears, the tip of the nose, the larynx, the trachea, the end of connecting bones, and in between the vertebrae that make up the backbone. In many places, cartilage has a supportive nature in the body, but in the vertebrae of the backbone it acts as a cushioning structure.

Human Muscles

In humans, there are three major types of muscle: **skeletal**, **smooth**, and **cardiac**. Each of these three types of muscles has unique functions within the human body.

Smooth muscles are found in the organs of the body. Smooth muscle is usually not under voluntary control. Smooth muscle consists of spindle-shaped cells that contain their own individual nuclei. The cells within the muscle also do not have any stripes or striations. Smooth muscles are found in the walls of the digestive tract, blood vessels, and other internal organs. Many smooth muscles can function in a unique manner by allowing electrical impulses to travel from one smooth muscle to another, without the aid of nervous stimulation.

Skeletal muscles are attached to the bones of the body and control their movement. Skeletal muscles control the voluntary actions of the body, such as writing, swinging a baseball bat, or walking. Skeletal muscles are made up of cells that are large, have more than one nuclei, and have stripes or striations going across the cell surface when viewed under a microscope.

Cardiac muscle is only found in the heart. Cardiac muscle, like smooth muscle, is made of cells that have the nuclei within the cytoplasm of each cell. The muscle tissue has striations like skeletal muscles. Cardiac muscle is an involuntary muscle that controls the beating of the heart.

Muscles and Bones

In the human body, the muscles and bones are connected in ways that permit movement of such structures as the arms and legs. **Tendons** are fibrous cords that attach muscles in the body to bones. The biceps and triceps muscles in the arm are attached to bones that permit the upward and downward curling action of the arm.

Bones are connected to other bones at movable joints by structures called **ligaments**. Ligaments are strong, elastic fibers found in such areas as the elbow joint and knee.

Practice Section 7-8

1. What best describes the skeletal structure of humans?
   a) endoskeleton c) interconnected ligaments
   b) exoskeleton d) none of these
2. Which chemical is found in the largest percentage in human bone?
   a) protein c) calcium phosphate
   b) sugar d) iron
3. What types of joints are found in the human skull?
   a) fixed joints c) freely movable joints
   b) slightly movable joints d) no joints
4. What part of the human body would you find cardiac muscle?
   a) digestive tract b) heart c) brain d) backbone
5. Tendons are fibrous cords that attach
   a) muscles to bone c) muscle to muscle
   b) bone to bone d) bone to organs
6. Muscle that cannot be consciously controlled is
   a) skeletal b) voluntary c) involuntary d) flexor

Lesson 7-9: Human Reproduction

Reproduction is the replication of another organism based on the genetic code transferred from one parent in **asexual reproduction** or from two parents in **sexual reproduction**. If you recall in the chapter on evolution, a species is considered biologically successful if it produces offspring that contain the genetic information of the parent or parents. The offspring produced from humans contains half of the genetic information from the male parent and half of the genetic information from the female parent. The human male and the human female contain specialized structures that enable the genetic information contained within each of their bodies to be united in the process of fertilization.

Human Reproductive Structures Male

The human male has two oval structures called **testes** which are contained inside a bag of skin called the **scrotum** that is suspended outside the body of the male. The testes are housed outside the body to keep them the necessary few degrees cooler than the body temperature, which ensures proper development of the **sperm**. The sperm contains the genetic information of the male and is manufactured in the testes. Each normal sperm contains 23 chromosomes. The testes also produce the male hormone **testosterone**. Connected to each testicle (singular of testes) is a long narrow tube called the **vas deferens**, which runs into the lower abdomen of the body and then loops back around down into the penis. Various glands connected to the vas deferens deposit fluids with the sperm as it travels to the penis for ejaculation from the body. **Seminal vesicles** secret a sugary mix to help nourish the sperm and the **prostate gland** secretes an alkaline solution to help protect the sperm from the acidic environment in the vagina. All of these fluids with the sperm create a mixture called ***semen***.

Male Reproductive Structures

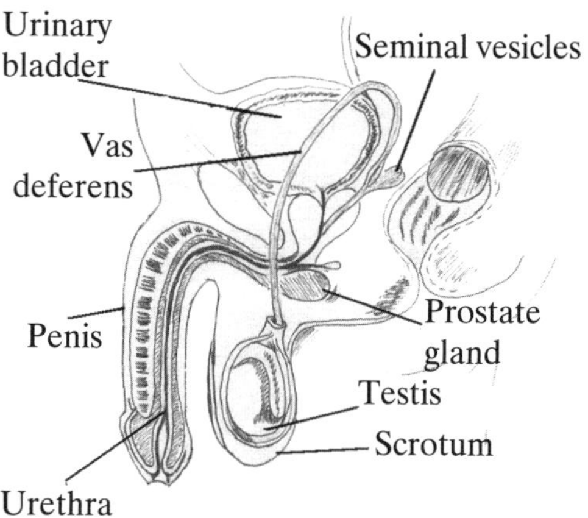

When the human male becomes sexually aroused, the penis tissue fills with blood, causing this structure to become elongated and erect. The erection opens the tube within the penis called the **urethra**, which allows the ejaculation of semen. A normal male ejaculation releases about 300 million sperm cells in the mix of about 3 ml of semen. As we learned earlier, the urethra in the penis also releases urine during excretion of waste. To prevent this from happening during the ejaculation of sperm, the exit to the bladder is shut.

Female

The female reproductive organs are located in the lower abdomen and are called the **ovaries**. The ovaries produce the **ova**, or eggs, that contain the genetic information of the female. Each female egg cell contains 23 chromosomes. The ovaries also produce the female hormones estrogen and progesterone. Each ovary is placed very close to the openings of the **fallopian tubes.** The fallopian tubes, or oviducts, are long ducts that open up into the womb, or **uterus**. The uterus is a thick-walled muscular organ where development of the human embryo takes place. The bottom of the uterus is a narrow muscular ring called the **cervix**. This muscular ring keeps the baby within the uterus during development. The cervix has a very small opening into the **vagina**, or female birth canal, that opens to the outside environment and receives the sperm for copulation.

The Female Menstrual Cycle

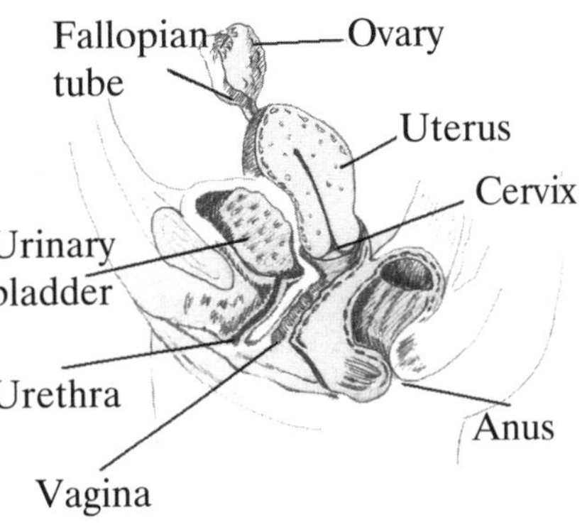

From the onset of puberty, the human male produces sperm on a daily basis almost until death. On the other hand, the human female is born with all the eggs she will ever have, about 450,000. These eggs are not yet mature and are stored in small clusters of cells called ***follicles*** in the ovaries. As a female reaches puberty her body will produce hormones that will cause these eggs to mature, usually one every 28 days throughout her life until menopause (around 50 years of age).

The menstrual cycle starts with the anterior pituitary gland secreting follicle-stimulating hormone (FSH) that causes the follicles in the ovaries to grow. As the egg inside matures, the follicle begins to secrete estrogen. The egg continues to grow and estrogen continues to be deposited into the bloodstream for a period of about 14 days. This higher level of estrogen in the blood causes the uterine lining to thicken and prepare for implantation of a possible fertilized egg. When these estrogen levels reach a certain critical level, luteinizing hormone (LH) is released from the anterior pituitary gland, which initiates the release of the egg from the follicle in a process called **ovulation**. In fact, many birth control pills work by blocking the release of LH in the female's body. The freed egg will then enter the small opening of the fallopian tube on its way to the uterus. The follicle, after releasing the egg, becomes filled with a yellow material called the ***corpus luteum***. The corpus luteum then begins to produce the hormones estrogen and progesterone that continue to prepare the females body for the possibility of a developing baby. After another 14 days, if the fertilized egg does not implant, the corpus luteum breaks down and the levels of estrogen and progesterone dramatically decrease. This, in turn, causes the extra blood and tissue connected to the uterine lining to break down and eventually slough off and be eliminated through the female's vagina. The release of this blood and tissue from the uterus is sometimes called the "period."

Fertilization

When the egg is being carried down the fallopian tube by millions of microscopic hairlike structures called cilia, it may come in contact with sperm. If a sperm cell meets the egg cell and their nuclei fuse, a diploid cell (full 46 chromosomes) is produced. This diploid cell is called a **zygote**. As the zygote continues to move down the fallopian tube it begins to divide by mitosis.

By the time the zygote reaches the uterus (usually a three-day journey) it has become a solid ball of many cells called a **morula**. This structure implants into the thickened wall of the uterus and begins to release its own hormones, which, in turn, prolongs the release of estrogen and progesterone from the corpus luteum. The implanted morula develops into a hollowed-out ball of cells called a **blastula**. The **gastrula** phase begins when the cells of the blastula begin to bend inward at the base. At the end of the gastrula phase there are three distinct layers of cells.

The three distinct layers of cells are:

- The **endoderm** (inner cells).
- The **mesoderm** (middle layer of cells).
- The **ectoderm** (outer cells).

••••••

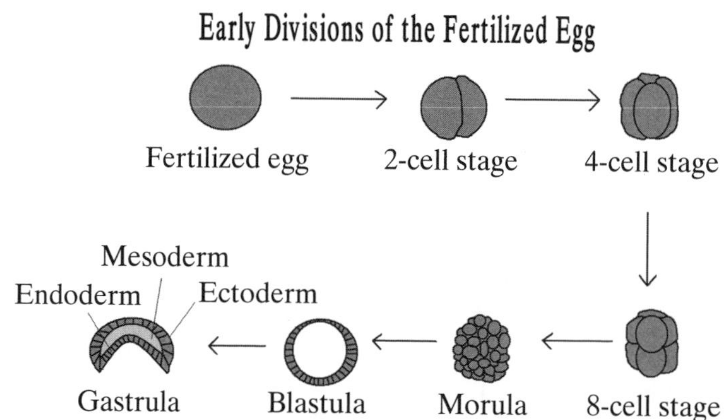

These layers will eventually show that these cells have become ***differentiated***.

In about eight weeks the developing cells are classified as a **fetus.** Membranes surround and protect the developing embryo in the uterus. The outermost membrane, called the **chorion**, forms the placenta. The embryo is attached by the **umbilical cord** to the placenta in the uterus. The **placenta** provides the connection point for the transfer of food and oxygen from the mother to the baby by way of the ambilical cord. This transfer is accomplished by the diffusion of these materials across the membranes; there is no mixing of blood between the mother and the baby. These membranes are also close enough for the diffusion of

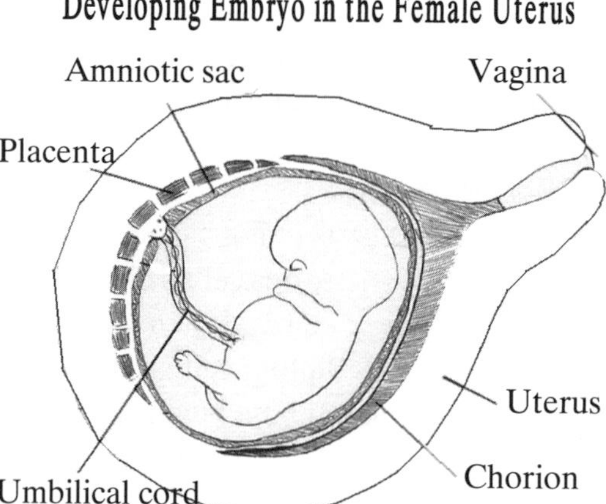

many possible hazardous chemicals, such as alcohol and nicotine. Surrounding the developing embryo is another layer called the **amnion** that is filled with amniotic fluid, which cushions any sudden jolts the embryo could experience inside the uterus during development.

Practice Section 7-9

1. What is the name of the structure that houses the human male testicles outside of the body?
   a) prostate gland b) vas deferens c) scrotum d) cervix
2. What is the name of the female reproductive structure that produces the sex gametes?
   a) ovary b) fallopian tube c) cervix d) vagina
3. Semen leaves the human male body from what structure?
   a) cervix b) urethra c) follicle d) morula
4. What is the normal genetic condition of a human zygote?
   a) monoploid b) haploid c) diploid d) A and B e) none of these
5. Each human gamete contains
   a) the diploid number of chromosomes
   b) 45 chromosomes
   c) 23 chromosomes
   d) 23 genes
6. In humans, the process of fertilization normally takes place in what structure?
   a) vagina b) oviduct c) uterus d) ovary

Lesson 7-10: Human Defense System

The human body encounters many dangers every day, such as attempting to walk across a busy street, driving in hazardous weather conditions, or possibly encountering job-related accidents. But it is the unseen perils to the body of pathogenic bacteria, fungi, and viruses that could be the most dangerous were it not for the human defense systems. The human defense system is all the cells, organs, and associated chemicals that protect the body against these foreign invaders.

The human body's first way to protect itself against the invasion of pathogens is to not allow them inside the body. The skin is an organ that is very good at this job. The skin not only provides a physical barrier to many foreign substances, but it also has chemicals on its surface that can destroy many of these invaders. For example, in a case when someone receives severe burns to the skin, the body loses the protection of this physical barrier. The most dangerous event a doctor treating a burn victim must be on guard against is the possibility of infection, because of the open pathway to the internal systems of the body caused by the burn. Unfortunately, even with the protection of the skin, many pathogens do, at times, make their way into the bodies of humans. There are openings to the internal parts of the body in the skin such as the eyes, nose, ears, reproductive organs, and mouth that become easier spots for the entry of these invaders. Once these pathogens are inside the body, other defense mechanisms kick in to combat the invaders.

The Lymphatic System

The lymphatic system is a series of vessels found throughout the human body that collects and returns fluids that have leaked from the bloodstream. **Lymph** is the name given to the collected fluid found in these vessels and has a makeup similar to blood plasma with the addition of white blood cells. There are large masses of lymph tissue found in various sections of these lymph vessels that contain large numbers of white blood cells that aid in the defense of the body. The lymphatic vessels are also found in the villi of the small intestines in the form of lacteals that absorbs fat from digested food.

White Blood Cells (Leukocytes)

Cell Type	Percentage of White Blood Cells in the Body	Function
Eosinophils	~5%	A type of white blood cell important in the allergic response mechanism in humans because they have antihistamine properties.
Neutrophils	~64%	A type of white blood cell that engulfs and then releases corrosive chemicals that kill all surrounding pathogens and the neutrophil cell itself. The formation of pus is usually a sign of these cells in action.
Monmocytes	~5%	The largest type of white blood cell in the body that engulfs and then digests pathogens.
Basophils	~1%	This type of white blood cell ingests bacteria and produces the chemicals heparine and histamine.
Lymphocytes	~25%	A type of white blood cell with a very large nucleus. This type of white blood cell produces antibodies against toxins produced by pathogens.

Table 7.5

Cellular Defenses

As mentioned, white blood cells are a critical component in the defense against pathogens that have entered the body. In the human body, there are normally about 8,000 white blood cells per milliliter of blood; this number will rise in the event of infection. Inside the human body there are five types of white blood cells that protect the body in different ways. Table 7.5 on page 236 shows these cells and their function.

The white blood cells are produced in the bone marrow, lymph nodes, and the spleen. Most of the white blood cells mentioned in Table 7.5 move randomly throughout the blood and organ tissues of the body looking to kill pathogens. The lymphocytes conduct a very organized approach to killing invaders within the body in what is called the human **immune response**.

The Immune Response

The human immune response is triggered by the release of an alarm chemical called ***interleukin-1*** by the macrophage white blood cells. This alarm chemical initiates the response of a type of lymphocyte called **helper T cells**. These helper T cells do not actively kill pathogens in the body but instead stimulate two more kinds of lymphocyte white blood cells called killer T cells and B cells to respond. Killer T cells patrol the blood and lymph fluid and attack and destroy pathogens. Killer T cells are able to recognize foreign substances if the receptor proteins found on their cell membrane match the pathogen. The body produces killer T cells with many different kinds of receptor proteins; the type of cell with the correct match to the pathogen will eventually be produced by the body in the largest quantities. The B cells produce a substance called an **antibody** that circulates in the blood and

Schematic of Immune Response

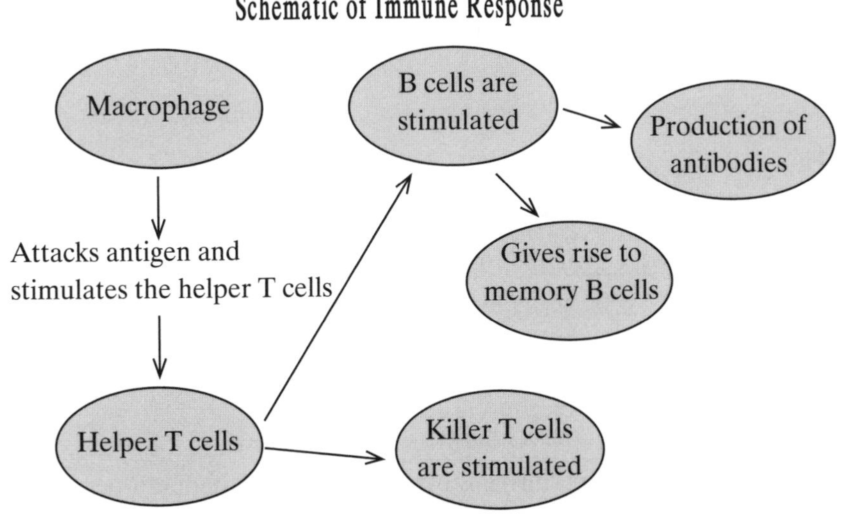

lymph until it attaches to a foreign particle, marking it for destruction by the T cells. Some antibodies can also block viruses from entering body cells. A very important function of some of the B cells in the body is to remember the pathogen that attacked the body. These memory B cells will quickly initiate the cellular defenses against a pathogen that has previously attacked the body. This quick response to a pathogen entering the body for the second time lessens the dangerous effects of this pathogen, sometimes so much so that you will not experience the sensation of feeling ill as your body fights off the invader. This is what a person being "immune" to a certain disease means.

Fever Defense

You can probably remember the last time you had a fever. Many times when individuals get a fever they reach for some type of drug to bring the fever down. Actually, in many cases (as long as the fever does not get too high), having a fever is helping your body to regain good health. When some pathogens are recognized by white blood cells, they alert the anterior hypothalamus of the brain to increase the normal body temperature. Raising the body's temperature has some positive effects in the battle against the pathogen. Many pathogenic organisms grow slower in high temperatures, allowing the body more time to build up defenses. Also, by raising the body temperature, the metabolic rate of the body is increased. This increases the rate of white cell production and speeds up the repair of damaged tissue. When the temperature of an adult begins to rise about the 103° F range, enzymes in the body can begin to denature. This is why very high fever should be brought down with medicine.

Practice Section 7-10

1. What is considered the human body's first line of defense?
   a) white blood cells b) hydrochloric acid c) skin d) lymphatic system
2. What white blood cells are found in the highest percentage within the human body?
   a) eosinophils b) monocytes c) neutrophils d) basophils
3. What cell within the body usually initiates the immune response?
   a) killer T cell b) B cell c) helper T cell d) antibody
4. A substance that triggers an immune response is a(n)
   a) antibody b) antigen c) mucus d) histamine
5. What is true about the human defense mechanism of raising the body temperature?
   a) The cerebellum increases the body temperature.
   b) Most pathogenic organisms reproduce slower at high temperatures.
   c) The number of white blood cells decreases.
   d) All of these.

6. What is the defensive function of having mucus membranes in most of the openings to the human body?
   a) Mucus prevents fluids from leaking out into the environment.
   b) Mucus traps possible pathogens before they can enter the body.
   c) Mucus keeps the body openings cool to prevent disease.
   d) Mucus digests harmful pathogens.

Lesson 7-11: Human Disease

A **disease** is any change within the body (other than injury) that disrupts homeostasis. Agents such as fungi, bacteria, or viruses can cause disease in humans. Any organism that can cause disease in humans is called a **pathogen.** Some materials found in the environment can also cause disease, such as the chemicals found in the smoke of cigarettes. There are also diseases caused from inheriting damaged genes at conception, causing such diseases as sickle-cell anemia.

In the past, many people believed that disease was caused by evil spirits, a curse that was placed upon you, or if you upset the gods. It was not until the mid-1800s that the work of scientists such as Louis Pasteur and Robert Koch showed the world that in many cases, disease was caused by organisms living in the environment, not evil spirits.

The Work of Louis Pasteur

Louis Pasteur (1822–1895) was a French chemist who had many breakthroughs in the understanding of disease and how it is spread. In his early research, he was hired by wine growers of France to discover why some wine turned sour and other wine did not. During his research Pasteur discovered that the wine that had soured had microorganisms living in it, while the "good" wine did not. Louis Pasteur also showed that by boiling the wine and then cooling it down, the microorganisms were killed. He called this heating and cooling process "pasteurization."

This breakthrough in the tests done on wine motivated Pasteur to show the same connection with people who had a disease. Pasteur compared the blood of healthy people to the blood of people who had a disease. The people's blood that had a disease was full of these microorganisms.

Louis Pasteur went on to discover vaccinations for cholera, anthrax, rabies, chicken pox, and diphtheria. His work also led to widespread changes in hospitals, where surgical equipment would now be boiled to kill any germs that might be on the instruments that could lead to disease in the patients.

Much of the work of Louis Pasteur would lead the way for another scientist, Robert Koch, to continue the study of disease and germs.

Robert Koch

Robert Koch (1843–1910) was a German scientist who, after reading some of the work of Louis Pasteur, began to research and conduct experiments concerning how microbes affect animals and people.

Robert Koch discovered that it is germs that cause infection in wounds. Koch stained these microbes with dye, which allowed him to photograph them under the microscope. This work led Koch to show that most diseases can be traced back to a particular microorganism. Koch was the first person to identify the organisms that cause the deadly diseases of tuberculosis, cholera, and anthrax.

Robert Koch formulated a specific, scientific series of steps for proving if a certain germ causes a certain disease. These steps have become known as **Koch's Postulates.**

Koch's Postulates

1. **The same microorganism is found in all people or animals that have the disease.**
2. **This microorganism must be isolated and grown in a lab setting.**
3. **When the microorganism that is grown in the lab is injected into a healthy animal, it must cause the disease.**
4. **The same microorganism should be obtained from the new diseased animal.**

••••••

Edward Jenner and Vaccination

You probably can remember going to the doctor at some point to receive a shot. Although getting this shot at the doctor's office was probably not something you were looking forward to, it was a very effective form of protection against disease called a vaccination. **Vaccination** is when either a dead or weakened form of a pathogen is injected into the body to initiate an immune response. Because the pathogen is either dead or weakened, the body is able to build up a natural immune response to this pathogen without the risk of getting the disease. Millions of lives have been saved since the advent of this miraculous type of disease prevention.

A country doctor by the name of Edward Jenner started the medical process of vaccination. During Jenner's time, millions of people around the world died every year from a disease called smallpox. By way of astute observation, Jenner noticed that many milkmaids only received a weak version of smallpox, not the more fatal type that most people in the surrounding area were getting. It was eventually realized that these milkmaids where not getting smallpox but a similar disease called cowpox from the cows they were milking. Cowpox produced blisters on the milkmaids' hands that cleared up in a few days with no other symptoms. The milkmaids who had the blisters never developed smallpox. Edward Jenner believed that something from the cowpox blisters was preventing these milkmaids from getting smallpox. Edward Jenner tested this idea by draining the pus from the blisters on the hands of a milkmaid and injecting it into a small boy named James Phipps. Later, Jenner injected smallpox into this same boy. James never developed the severe form of smallpox. The world had experienced the first case of vaccination preventing disease.

Over the many years that followed, millions of people around the world would be vaccinated against smallpox and other diseases. In fact, vaccination against smallpox was so successful that there has not been a new case of smallpox in the entire world since 1972.

Practice Section 7-11

1. Which of the following would not be considered a pathogen?
   a) bacteria b) pollen c) virus d) fungi
2. What is pasteurization?
   a) The boiling and cooling of a substance to kill possible pathogens.
   b) The method of separating substances in a mixture by evaporation of a liquid and the following condensation of its vapor.
   c) The reaction of an acid with a base to produce salt and water.
   d) The phase change of a substance from a solid directly to a vapor.
3. Louis Pasteur is responsible for which of the following reforms?
   a) The present-day classification system of using binomial nomenclature names to identify organisms.
   b) All surgical equipment must be boiled to kill germs before surgery.
   c) The vaccination of people to prevent disease.
   d) The specific scientific steps taken to prove a specific germ causes a specific disease.
4. Which of the following is not part of Koch's postulates?
   a) The same microorganism must be found in all people or animals that have a particular disease.
   b) The microorganism must be isolated and grown in a lab setting.
   c) The microorganism must be unicellular in structure.
   d) When a particular disease causing microorganism is grown in the lab and then injected into a healthy animal, it must cause the disease.
5. Which scientist is credited with the development of medical vaccinations?
   a) Robert Koch c) Edward Jenner
   b) Charles Darwin d) William Harvey

Terms From Chapter 7

adrenal gland	heart	prostate gland
alveoli	hemoglobin	receptor sites
antibody	hormone	rectum
arteries	hypothalamus	red blood cells
asexual reproduction	immune response	scrotum
axon	Islets of Langerhans	semen
bile	kidneys	seminal vesicles
bones	large intestine	sexual reproduction
brain	ligaments	small intestine
capillaries	lungs	sperm
cerebellum	lymph	stomach
cerebrum	medulla	synapse
cervix	myelin sheath	tendons
chime	negative feedback	terminal branches
cyton	nephron	testes
dendrites	neuron	testosterone
diaphragm	neurotransmitters	thymus gland
disease	oral cavity	thyroid gland
emulsification	ovaries	trachea
endoskeleton	ovulation	urethra
epiglottis	parathyroid gland	uterus
esophagus	pathogen	vagina
estrogen	peristalsis	vas deferens
exoskeleton	pituitary gland	veins
fallopian tubes	placenta	villi
fetus	platelets	white blood cells
follicles	positive feedback	zygote
	progesterone	

Chapter 7: Exam

Matching Column for Human Digestive Organs

Match the digestive organ with its appropriate function. You may use an answer choice more than once or not at all.

Human Digestive Organs

1. stomach
2. large intestine
3. small intestine
4. liver
5. pancreas
6. rectum
7. esophagus
8. oral cavity

Functions

a. This organ absorbs most of the water from the undigested food.

b. This organ secretes the chemical enzymes of amylase, protease, and lipase.

c. This is a storage site for feces before being egested from the body.

d. This tube structure transports food from the oral cavity to the stomach.

e. This organ is the section of the alimentary canal where most of the food is absorbed into the blood.

f. The structure where mechanical digestion of food first occurs.

g. This structure is the site where the chemical breakdown of proteins first occurs.

h. This organ secretes the chemical bile, which is used to emulsify fats.

Matching Column for the Human Nerve Cell

Match the part of the human nerve cell with its appropriate function. You may use an answer choice more than once or not at all.

Parts of the Neuron

9. cyton
10. dendrite
11. axon
12. myelin sheath
13. terminal branches

Function

a. A fatty substance that covers the axon of the nerve cell and speeds transmission of the impulse.

b. Receives the stimuli sent from another nerve or the outside environment.

c. The body of the nerve cell that contains the organelles.

d. Transmits the nerve impulse to other nerves, organs, and glands in the body.

e. The long, thin section of the nerve cell where the impulse is transmitted across.

Multiple Choice

14. A heart attack is usually triggered when there is a blockage of
    a) the veins in the leg
    b) the blood vessels in the brain
    c) the coronary arteries
    d) the blood vessels in the kidney

15. White blood cells are also called
    a) erythrocytes b) leukocytes c) fibrin d) lymph

16. A small flap of skin that covers the trachea when food or liquid is swallowed to prevent choking is called the
    a) tonsils b) epiglottis c) alveoli d) bronchioles

17. What section of the brain is responsible for controlling basic life processes such as breathing and the beating of the heart?
    a) medulla b) cerebellum c) cerebrum d) none of these

18. Which of the following glands has influence over most of the other glands in the body and because of this is sometimes referred to as the "master gland"?
    a) thymus gland b) thyroid gland c) adrenal gland d) pituitary gland

19. What is the name of the human male hormone produced in the testes?
    a) estrogen b) progesterone c) testosterone d) adrenaline

20. In humans, at the end of the auditory canal there is a thin membrane called the
    a) retina b) tympanum c) cochlea d) stirrup

21. Bones are connected to other bones in the human body by strong, elastic, fibers called
    a) tendons b) ligaments c) cartilage d) muscles

Short Response

22. Describe the pathway of the sperm from the production point in the testes to its ejaculation point at the penis. Mention all glands that are involved in this process.

23. Explain the human immune system response and be sure to include the following terms in your explanation: *interleukin-1*, *helper T cells*, *killer T cells*, *B cells*, and *antibodies*.

24. Explain the steps of Robert Koch's postulates. How have these postulates improved the study of disease in humans?

25. Explain the process of human breathing. What function does the diaphragm have in this process?

26. What is "heartburn"? How could the muscle called the cardiac sphincter be involved in this condition?

Answer Key
Answers Explained Section 7-1

1. **B** is the correct answer because the chemical digestion of proteins begins in the stomach.

   A is not correct because proteins are mechanically digested in the oral cavity or mouth, but no chemical digestion of proteins occurs here.

   C is not correct because chemical digestion of proteins does occur in the small intestines following chemical digestion in the stomach.

   D is not correct because no chemical digestion occurs in the large intestines.

2. **B** is the correct answer. The enzyme amylase is used to aid in the chemical digestion of carbohydrates.

   A is not correct because the protease enzymes aid in chemical digestion of proteins.

   C is not correct because the lipase enzymes aid in the chemical digestion of lipids.

   D is not correct because water is not chemically digested in the human body.

3. **C** is the correct answer because the wavelike muscle contractions that bring food down the esophagus into the stomach is called peristalsis.

   A is not correct because pepsin is the enzyme found in the stomach that aids in protein digestion.

   B is not correct because dehydration synthesis is the chemical process of creating large complex molecules from smaller molecules with the removal of water.

   D is not correct because angina is the medical condition where coronary blood vessels are partially occluded, causing reduced oxygen flow to some heart tissue.

4. **D** is the correct answer because no chemical digestion occurs in the large intestine. The large intestine removes water from the materials that enter.

   A is not correct because mechanical digestion of all foods and the chemical digestion of some carbohydrates occur in the oral cavity.

   B is not correct because mechanical digestion of food continues in the stomach and the chemical digestion of proteins.

   C is not correct because the chemical digestion of carbohydrates, proteins, and lipids occurs in the small intestine.

5. **B** is the correct answer because emulsification is the physical breakdown of large lipid molecules into smaller lipid molecules in the small intestine by the chemical bile.

A is not correct because the digestion of protein in the stomach is achieved by protease enzymes.

C is not correct because the digestion of carbohydrates in the mouth is achieved by amylase enzymes.

D is not correct because the removal of undigested food from the rectum is called elimination.

6. **D** is the correct answer because all three enzymes—amylase, protease, and lipase—are released into the small intestines from the pancreas.

7. **D** is the correct answer because the specific protease enzyme that is used to break down the organic molecule protein is called pepsin.

   A is not correct because lipids are broken down by lipases.

   B is not correct because carbohydrates are broken down by amylases.

   C is not correct because water is not chemically broken down in the stomach.

8. **B** is correct because the breakdown of lipids into their subunits of glycerol and fatty acid molecules is an example of chemical digestion.

   A is not correct because the mechanical digestion of a lipid would be breaking down a large lipid molecule into a smaller lipid molecule.

   C is not correct because dehydration synthesis is the chemical processes used to create larger molecules, not the chemical process used to break these molecules down into smaller molecules.

9. **A** is the correct answer because villi are the microscopic structures found along the inner tract of the small intestine that increase surface area for food absorption.

   B is not correct because cilia are short rod-like structures that extend from the cell membrane of some one-celled organisms that aid in locomotion.

   C is not correct because flagella are long whiplike structures that extend from the cell membrane of some one-celled organisms that aid in locomotion.

   D is not correct because capillaries are small blood vessels that transport food and oxygen to cells in the body.

10. **B** is the correct answer because the epiglottis is the structure that covers the opening to the wind pipe to prevent food from entering.

    A is not correct because the larynx is the voice box.

    C is not correct because the pharynx is the posterior section of the oral cavity.

D is not correct because villi are structures found in the small intestine that increase surface area for food absorption.

11. Physical and chemical digestion begins in the human mouth. The teeth grind and soften food into smaller pieces to start the process of physical digestion. The enzyme amylase begins the chemical digestion of carbohydrates in the mouth also.

12. In humans the small intestines completes the chemical digestion of food by releasing enzymes that break down carbohydrates, proteins, and lipids. The chemical bile is also released in the small intestine. Bile helps break the fat globules into small enough pieces for chemical digestion. Most of the materials that have been chemically broken down in the small intestines are also absorbed by the small intestines into the blood stream. The large intestine on the other hand absorbs water back from the chemically digested food coming from the small intestines and with the help of symbiotic bacteria produces some important vitamins.

13. The stomach is protected from acids, used during digestion, by a thick mucus lining. An ulcer is a condition where the stomach lining has been broken down, usually by a type of bacteria. This means that when an ulcer has formed in the stomach, or sometimes the small intestines, the acid is eating away these structures.

Answers Explained Section 7-2

1. **A** is the correct answer because the function of red blood cells is to transport oxygen to the cells of the body.

   B is not correct because dissolved food is transported in the body by means of the blood plasma.

   C is not correct because hormones are transported in the body by means of the blood plasma.

   D is not correct because antibodies are transported in the body by means of the blood plasma.

2. **B** is the correct answer because the blood plasma is almost 90% water.

   A is not correct because alcohol is a poison to the body and would only be found in the plasma after someone had ingested this substance from the environment.

   C is not correct because sodium chloride is found in very small levels in the blood plasma.

   D is not correct because hormones are found in very small levels in the blood plasma.

3. **C** is the correct answer because platelets are responsible for the clotting of blood.

   A is not correct because white blood cells attack pathogens and infected body cells in defense of the body.

   B is not correct because red blood cells transport oxygen to the cells of the body.

   D is not correct because antibodies are part of the body's defense mechanism against invaders.

4. **D** is the correct answer because the left ventricle of the human heart pumps oxygenated blood to all parts of the body.

   A is not correct because the right atrium receives deoxygenated blood from the body and sends this blood to the right ventricle.

   B is not correct because the left atrium receives oxygenated blood from lungs and sends this blood to the left ventricle.

   C is not correct because the right ventricle receives deoxygenated blood from the right atrium and pumps this blood to the lungs.

5. **B** is the correct answer because the veins are the blood vessels that contain valves and send blood toward the heart from the extremities of the body.

   A is not correct because arteries are thick walled blood vessels that send blood under high pressure away from the heart to the extremities of the body.

   C is not correct because capillaries are very small blood vessels found throughout the body that are the location where most diffusion occurs between the blood and the cells of the body.

   D is not correct because lymph tubes transport the lymph fluid back into the blood vessels from places where this fluid has leaked out into spaces in the body.

6. **B** is the correct answer because thrombin is produced from the platelets to initiate the clotting of blood.

   A is not correct because lymph is not an enzyme.

   C is not correct because amylase is an enzyme that breaks down complex sugars.

   D is not correct because protease is an enzyme that breaks down proteins.

7. **C** is the correct answer because the most diffusion of materials between the blood and the blood vessels of the body occurs in the capillaries.

A and B are not correct because very little diffusion can occur between these vessels and cells in the body because of the thickness of the artery and vein vessels.

8. **A** is the correct answer because the substance that gives blood its red color is the chemical hemoglobin. Hemoglobin is reddish because it contains the element iron.

   B is not correct because sodium chloride is found in small amounts in the blood plasma and has little to no color.

   C is not correct because lipase is the name of enzyme found throughout the human body that aid in lipid digestion.

   D is not correct because chlorophyll is the green pigment found in photosynthetic organisms used to help trap sunlight for the food-making process.

9. **C** is correct because the biological name leukocytes refers to a class of white blood cells found in the body.

   A is not correct because platelets are responsible for blood clotting.

   B is not correct because red blood cells are responsible for the transport of oxygen and other gases in the body.

   D is not correct because lymph tubes transport the liquid called lymph throughout the body.

10. **C** is the correct answer because pulmonary circulation is the transport of blood between the heart and the lungs in the human body.

    A is not correct because blood flow between the heart and the liver is called portal circulation.

    B is not correct because blood flow between the heart and the kidneys is called renal circulation.

    D is not correct because blood flow between the heart and the brain is called cerebral circulation.

11. The blood coming from the body enters the human heart via the superior vena cava. The superior vena cava brings the blood into the first chamber of the heart called the right atrium. The blood then is pumped from the right atrium to the right ventricle. The blood is then pumped from the right ventricle out of the heart to the lungs through the pulmonary artery. Once the blood has circulated through the lungs and has become oxygenated, it returns to the heart via the pulmonary veins. The pulmonary veins bring the blood to the next chamber of the heart the left atrium. The blood is then pumped from the left atrium to the final chamber of the heart the left ventricle. The blood is pumped out of the left ventricle by way of the aorta.

12. Arteries are thick-walled flexible vessels that carry blood under high pressure (form the beating heart muscle) from the heart outward to the body. Veins are thin-walled tubes that bring the blood under low pressure from all the parts of the body back to the heart.
13. The lymphatic system is a series of tiny tubes located throughout the body that help collect all the plasma fluid that leaks out of the blood vessels during transport. The lymphatic system also contains many oval shaped structures called lymph nodes. These lymph nodes are packed with white blood cells and antibodies and act as another defense mechanisms for the body.

Answers Explained Section 7-3

1. **B** is the correct answer because the mechanical process of bringing air into the lungs is called breathing.

   A is not correct because respiration is the chemical oxidation of food to release the stored energy in the bonds of these chemicals.

   C is not correct because dehydration synthesis is the creation of large molecules by joining smaller molecules with the removal of water.

   D is not correct because oxidation is the chemical process where a molecule loses electrons.

2. **C** is the correct answer because the medical term for the human windpipe is the trachea.

   A is not correct because epiglottis is the flap of skin that covers the trachea during the swallowing of food or liquid.

   B is not correct because esophagus is the tube that connects the pharynx to the stomach.

   D is not correct because larynx is the term for the human voice box.

3. **D** is the correct answer because the location of gas exchange in the human body during the process of breathing is the alveoli in the lungs.

   A is not correct because trachea is the tube that brings gases from the oral cavity down into the bronchi tubes of the lungs. In the trachea, little to no gas exchange occurs between this tube and the cells of the body.

   B is not correct because bronchi are the tubes that branch from the trachea into each lung in the chest cavity of the human body. In the bronchi, little to no gas exchange occurs between these tubes and the cells of the body.

   C is not correct because bronchioles are the tubes that branch from the bronchi tubes into the inner areas of each lung. In the bronchioles, little to no gas exchange occurs between these tubes and the cells of the body.

4. **C** is the correct answer because the gases in the lungs are exchanged with the capillary blood vessels. The thin walls of the capillaries allow easy gas exchange between these vessels and the body cells.

   A and **B** are not correct because arteries and veins are not present in large numbers in the lung tissue. Arteries and veins also have much thicker vessel walls that do not allow the easy diffusion of gases.

   D is not correct because lymph tubes are not used for the movement or diffusion of gases in the body.

5. **A** is the correct answer because when the diaphragm muscle is in a downward curved position, air will enter the lungs. When the diaphragm is this downward position in the lower abdomen, it creates more space in the lung cavity that in turn reduces air pressure causing air from the environment to flood into the lungs.

   B is not correct because air will exit the lungs when the diaphragm is in the upward position, causing air pressure to increase and forcing these gases out of the body.

   C is not correct because air is always moving when the diaphragm is actively bending.

6. **C** is the correct answer because the chemical hemoglobin allows red blood cells to hold more oxygen.

   A is not correct because infections are fought by the white blood cells in the body.

   B is not correct because enzymes bind with food molecules in the process of digestion.

   D is not correct because worn-out red blood cells are replaced in the liver.

7. The diaphragm is a sheet of skeletal muscle located below the lungs that aids in the act of breathing. When someone inhales, the diaphragm contracts and this muscle flattens causing the diaphragm to move lower into the chest cavity. This action causes the air pressure inside the lungs to decrease, causing the gases with higher air pressure outside of the body to rush in (breathing in). When someone exhales gases, the diaphragm relaxes and this brings the chest muscles back to the normal position, and the diaphragm higher up the chest cavity. Now the gases inside the lungs are under higher pressure than the gases outside the body. This action causes the air in the lungs to rush outside the body (breathing out).

8. Oxygen is brought into the lungs through the mechanical process of breathing. The blood is the transport system for oxygen in the body. This means that the

oxygen in the lungs must be transferred to the blood. The alveoli in the lungs are covered in mass nets of capillaries, it is at this point where the lungs truly meet the circulatory system and the transfer of oxygen occurs into the blood. The blood contains the chemical compound hemoglobin which allows it too hold large quantities of oxygen for transport. The blood vessels transport the oxygenated blood throughout the parts of the body.

9. Breathing is a mechanical process where the diaphragm muscle contracts and expands causing the intake and release of atmospheric gases. Respiration is a chemical process where the oxygen that is acquired during breathing is used to release the energy trapped inside ingested food molecules.

Answers Explained Section 7-4

1. **B** is the correct answer because excretion is the removal of waste from the process of cell metabolism.

   A is not correct because egestion is the removal of waste, which is mostly undigested food, from the anus.

   C is not correct because synthesis is the creation of large, more complex molecules from smaller, simpler molecules.

   D is not correct because deamination is the removal of an amino group from an amino acid molecule.

2. **B** is correct because nitrogenous wastes are produced from the breakdown of proteins.

   A is not correct because carbohydrates are made up of simple sugars.

   C is not correct because lipids are made of glycerol and fatty acid molecules.

   D is not correct because simple sugars connected in chemical chains create polysaccharides.

3. **B** is the correct answer because the kidney filters the blood of cellular waste products.

   A is not correct because the liver breaks down worn out red blood cells, and also breaks down certain chemicals in the blood before they are sent to the kidney for removal.

   C is not correct because the large intestine removes water from the undigested food before egestion.

   D is not correct because the pancreas releases important enzymes into the small intestines for the digestion of food.

4. **B** is the correct answer. The nephron is the functional unit of the kidney. This means that the nephron is the actual structure in the kidney that filters the blood of waste.

A is not correct because the urethra is the tube in the male and female body through which urine leaves the body.

C is not correct because the ureter is the tube that sends the urine, filtered from the blood in the kidney, to the urinary bladder for storage.

D is not correct because dendrite is the end of the nerve cell or neuron that collects stimuli from the body or the environment.

5. **D** is the correct answer because all three chemicals (water, urea, and uric acid) are found in the urine.

6. **C** is the correct answer because the Loop of Henle is the section of the nephron where food is reabsorbed back into the bloodstream.

   A is not correct because the glomerulus is a collection of capillaries in the kidney where waste leaves the bloodstream and enters the nephron.

   B is not correct because Bowman's capsule is the section of the nephron that collects the waste from the bloodstream in the kidney.

7. The kidney is an organ that filters the blood of nitrogenous waste products. Water is recycled back into the blood, during this filtering process in the kidney, enabling the body to conserve large amounts of fluid.

8. A nephron is the functional unit of the kidney. The kidney organ is composed of millions of tiny filters called nephrons. Each nephron is a microscopic tube about 3 cm long that filters waste from the blood. A ball of capillaries, called the *glomerulus*, inside the nephron, squeezes water, food, salt and urea from the blood into the collecting area called the *Bowman's capsule.* All these materials move from the Bowman's capsule into a coiled tube (Loop of Henle) where some of the salts, food, and most of the water are reabsorbed back into the bloodstream. The urea, some salt, and some water remain in the tube and are eventually excreted from the body during the process of urination.

9. Egestion is the removal of fecal matter from the anus. Fecal matter consists of undigested food that has not entered the bloodstream. Excretion is the removal of waste productes (principally urea) from digested food during the process of urination.

Answers Explained Section 7-5

1. **D** is the correct answer because the dendrites are the sections of the neuron that receive impulses sent from other nerve cells or stimuli from the environment.

   A is not correct because the cyton is the body of the neuron that contains the organelles necessary for the life functions.

B is not correct because the myelin sheath covers the axon of some neurons and increases the speed of the impulse along the nerve.

C is not correct because the terminal branches transmit the nerve impulse from the neuron to another nerve, gland, or muscle in the body.

2. **B** is the correct answer because the chemicals responsible for the creation of the chemical charge that produces the impulse are sodium and potassium.

3. **C** is the correct answer because the strength of a stimulus from the environment is determined by both the number of impulses sent per second down the neuron and the number of neurons sending this signal from the stimulus.

4. **C** is the correct answer because neurotransmitters are released from one neuron and collected by the neuron on the other side of the synapse.

   A is not correct because the impulse is electrical in nature on the axon but this electrical signal does not cross the synapse.

   B is not correct because none of the neurons in the body are connected; they all are separated by small spaces called synapses.

   D is not correct. See answers **A** and **B**.

5. **A** is the correct answer because the sensory neurons send impulses from stimuli collected from the environment to the central nervous system.

   B is not correct because motor neurons send impulses from the central nervous system to glands or muscles in the body.

   C is not correct because interneurons connect sensory neurons to motor neurons in the central nervous system.

   D is not correct because neurotransmitters are chemicals sent to continue the nerve impulse from one neuron to another across the synapse.

6. **B** is the correct answer because the part of the brain called the cerebellum is responsible for coordination and balance of the body.

   A is not correct because cerebrum is the part of the brain responsible for higher-level thought and memory.

   C is not correct because medulla is the part of the brain that controls the functioning of organs, such as the heart and digestive system.

   D is not correct because the spinal cord is the part of the central nervous system that controls reflexes and the relay of impulses between the peripheral nervous system and central nervous system.

7. A reflex is an involuntary, automatic response to a given stimulus. Many normal body functions are controlled by reflexes, including blinking, sneezing, breathing, and heartbeat. The pathway that a nerve impulse travels in a reflex

action is called the reflex arc. The reflex arc starts when a stimulus from the environment or the body triggers a sensory neuron. The nerve impulse travels up the sensory neuron to the inter-neurons in the central nervous system (brain or spinal cord). The inter-neurons transmit the nerve impulse to a motor neuron, which will bring the signal to an effector (organ, muscle or gland) in the body.

8. The autonomic nervous system consists of the sympathetic and parasympathetic systems. These two systems send nerve signals to the same parts of the body but causes opposite effects at these locations. The sympathetic nerves increase heart rate, and release chemicals that prepare the body for action or movement. The parasympathetic nerves decrease the heart rate and calm the body by releasing chemicals that slow breathing.
9. The cerebrum is made up of gray matter and white matter. The outside of the cerebrum is called the cortex tissue and contains what is sometimes called gray matter. Gray matter consists mostly of cyton portions of the nerve cells. The inner section of the brain contains white matter and consists primarily of the axon portion of the nerve cells.

Answers Explained Section 7-6

1. **C** is the correct answer because the salivary gland is not part of the endocrine system. Glands that are part of the endocrine system must be ductless and secrete their hormones directly into the bloodstream. The salivary gland has ducts and does not secrete its fluids into the bloodstream.

   A, **B**, and **D** are not correct because the pituitary gland, adrenal gland, and thymus gland are all ductless glands that secrete their hormones into the bloodstream and, therefore, are part of the endocrine system.
2. **C** is the correct answer because homeostasis is controlled in the human body by both the nervous and endocrine system.
3. **A** is the correct answer because the pancreas secretes the chemical insulin.

   B is not correct because the liver secretes the chemical bile used for lipid digestion.

   C is not correct because the stomach secretes gastric digestive enzymes, such as pepsin.

   D is not correct because the brain contains a few structures, such as the pituitary gland and the pineal gland, that secrete various chemicals for coordination and control of body functions.
4. **B** is the correct answer because the hormone adrenaline is secreted during stressful situations. This hormone increases heart rate and breathing.

A is not correct because thyroxin is a hormone secreted by the thyroid gland the controls proper cell metabolism.

C is not correct because oxytocin is a chemical secreted by the pituitary gland and hypothalamus that promotes labour during pregnancy and the production of milk from the mammary glands.

D is not correct because TSH is secreted by the pituitary gland and stimulates the release of thyroxin from the thyroid gland.

5. **A** is the correct answer because hormones are transported through the body by the bloodstream.

   B is not correct because the lymphatic ducts are used to collect the runoff from the bloodstream called lymph.

   C is not correct because glands that contain tubes are part of the exocrine system not the endocrine system.

   D is not correct because the bones structure can receive hormones and respond to them, but the bone structure is not used to transport them around the body.

6. **B** is the correct answer because a negative feedback cycle is when a hormone causes a target tissue to release less of the chemical causing the initial action in the body.

   A is not correct because a positive feedback cycle is when a hormone causes a target tissue to release more of the chemical causing the initial action in the body.

   C is not correct because cyclosis is the circular movement of the cytoplasm that occurs in some plants cells.

   D is not correct because an aneurysm is a bulge in a blood vessel that weakens the wall of that blood vessel. This is a dangerous condition because the blood vessel in this weakened condition can burst, causing internal bleeding.

7. A hormone is a chemical secreted by cells in one part of the body and affect cells in other parts of the body. In humans hormones are usually transported in the body by the blood. Usually very small amounts of hormone are needed to produce the desired reaction.

8. The thyroid gland produces the hormone called thyroxin. The chemical thyroxin regulates the rate of metabolism in the body. This chemical is essential for normal mental and physical development.

9. The pituitary gland is connected to the hypothalamus at the bottom of the brain. The pituitary gland is many times called the master gland because its

hormones can affect so many other glands within the body. Some of the major hormones secreted by the pituitary gland include the growth hormone, thyroid stimulating hormone, oxytocin, and the follicle stimulating hormone.

10. In human females the ovaries produces the hormone called estrogen. Estrogen is the chemical responsible for female secondary sex characteristics such as breast enlargement, and narrowing of the hips. The hormone estrogen and progesterone also help to build of the uterine lining for pregnancy. In human males the testes produce the hormone called testosterone. Testosterone is responsible for the male secondary sex characteristics such as facial hair, the deepening of the voice, and broad shoulders. Testosterone is also important in the production of sperm in the testes.

Answers Explained Section 7-7

1. **B** is the correct answer because the ears collect the stimulus of sound waves from the environment.

   A is not correct because the eyes collect the stimuli of light.

   C is not correct because the nose collects the stimuli of chemicals.

   D is not correct because the tongue also collects the stimuli of chemicals.

2. **B** is the correct answer because the olfactory nerve receives the impulses from stimuli picked up by receptors in the nose.

   A is not correct because taste buds are found on the surface of the tongue.

   C is not correct because cochlea is a structure in the inner ear responsible for the collection of sound waves.

   D is not correct because auditory canal is the opening from the environment into the inner ear.

3. **C** is the correct answer because your sense of taste will also be affected if your nose is stuffy. The nose and tongue pick up chemicals in the environment and the brain interprets the signal sent by them. The tongue and nose work together in collecting these chemicals for taste.

   A is not correct because sense of sight would not usually be affected by a stuffy nose.

   B is not correct because sense of touch would not usually be affected by a stuffy nose.

   D is not correct because sense of hearing would not usually be affected by a stuffy nose.

4. **C** is the correct answer because the inner ear structure called the semicircular canals aid the brain in balancing the body.

A is not correct because the part of the inner ear responsible for sound reception is the cochlea.

B is not correct because the tympanum is the name for the ear drum location at the edge of the inner ear.

D is not correct because the anvil is one of the three tiny bones in the inner ear that picks up the sound vibrations entering the ear canal.

5. **D** is the correct answer because the brain does not contain sensory neurons for touch.

   A, **B**, and **C** the fingers, toes, and face all contain many sensory neurons for the stimuli of touch.

6. **B** is the correct answer because the structures called cones are responsible for distinguishing the different colors of light that enter the human eye.

   A is not correct because rods are the structures present in the eye that aid the eye in collecting light of varying intensities.

   C is not correct because the pupil is the part of the eye that regulates the amount of light that enters into the eye from the environment.

   D is not correct because the lens is the part of the eye that helps focus the incoming light from the environment.

7. The tympanum is a thin membrane sometimes called the eardrum located at the end of the auditory canal in the inner ear.

8. The iris controls the opening of light into the inner eye. The pupil is the opening caused by the movement of the iris. The cornea focuses light that has entered the eye from the environment. Then photoreceptors located on the retina of the inner eye convert the light energy into nerve impulses that will be sent to the brain by the way of the optic nerve.

9. All nerve impulses (no matter the strength of the stimulus) that travel from the sensory neuron to the central nervous system are of the same strength. The way the body interprets a higher magnitude stimulus from the environment is by doing two things; sending the nerve signal across many more neurons, and increasing the frequency that the signal is transmitted across those neurons.

Answers Explained Section 7-8

1. **A** is the correct answer because humans have endoskeletons. Endoskeletons are skeletons found inside the tissue layer of the body.

   B is not correct because exoskeletons are skeletons found on the outside of the body. Animals such as insects have exoskeletons.

   C is not correct because the ligaments of the human body are used to connect bones to other bones.

2. **C** is the correct answer because the chemical found in the highest percentage in human bones is calcium phosphate. Calcium phosphate makes up about 70% of the bone structure.

   A is not correct because protein is the second most common chemical in the makeup of bones. Protein makes up about 30% of the bone structure.

   B is not correct because sugar feeds the live cells in the interior matrix of the bone structure.

   D is not correct because iron is an important chemical in the makeup of certain cells in the body, such as red blood cells.

3. **A** is the correct answer because the skull contains what is called fixed joints. These joints in the skull are present but do not move.

   B is not correct because slightly movable joints are joints with some ability of movement. The space between the vertebrate of the backbone contains slightly movable joints.

   C is not correct because freely movable joints are joints that have considerable movement. The joints between the arm bones and legs bones contain freely movable joints.

   D is not correct because there are joints in the skull.

4. **B** is the correct answer because the only place in the human body in which cardiac muscle would be found is in the heart.

   A is not correct because the digestive tract contains smooth muscles.

   C is not correct because the brain does not contain muscle.

   D is not correct because the backbone contains cartilage in the spaces between the vertebrate.

5. **A** is the correct answer because tendons connect muscles to bone in the body.

   B is not correct because ligaments connect bone to bone in the skeletal structure.

   C is not correct because muscles are connected to bone in the body.

   D is not correct because organs are not connected directly to bone in the human body.

6. **C** is the correct answer because the muscles in the body that cannot be consciously controlled are called involuntary muscles.

   A, **B**, and **D** are not correct because skeletal muscles, voluntary muscles, and flexor muscles are all under voluntary control.

Answers Explained Section 7-9

1. **C** is the correct answer because the scrotum houses the male testicles outside of the body. The scrotum keeps the testicles at the proper temperature.

   A is not correct because the prostate gland secretes an alkaline solution that protects the semen inside the acidic environment of the female vagina.

   B is not correct because the vas deferens are the tubes that connect the testicles to the urethra in the penis. These tubes are where the sperm travel from the testicles to the urethra during ejaculation.

   D is not correct because the cervix is the narrow passageway that connects the vagina to the uterus in the female reproductive system.

2. **A** is the correct answer because the female reproductive structure that produces the sex gametes is called the ovary. The ovary produces the female eggs.

   B is not correct because the fallopian tubes are the passageways where the egg and sperm meet before the fertilization egg moves into the uterus.

   C is not correct because the cervix is the narrow passageway that connects the vagina to the uterus in the female reproductive system.

   D is not correct because the vagina is the opening to the female reproductive organs. The male's penis is inserted in the vagina during copulation for discharge of sperm.

3. **B** is the correct answer because the semen leaves the human male body by way of a tube in the penis called the urethra.

   A is not correct because the cervix is the narrow passageway that connects the vagina to the uterus in the female reproductive system.

   C is not correct because the follicle is a cavity in the ovary that contains the developing egg.

   D is not correct because morula is the name given to the developing egg in the female body once it has become a solid mass of cells and implants into the uterine lining.

4. **C** is the correct answer because the normal genetic condition of the zygote is diploid. In humans, a diploid zygote means there are 46 chromosomes present.

   A is not correct because monoploid and haploid both mean the genetic condition for a human cell would be 23 chromosomes. Only the male and female sex cells contain 23 chromosomes, the zygote contains 46.

5. **C** is the correct answer because each human gamete, or sex cell, contains 23 chromosomes.

A is not correct because the diploid number of chromosomes (46) is found in the normal body cells of the human body.

B is not correct because 45 chromosomes would be present in a body cell that has experienced a chromosome mutation.

D is not correct because genes make up the chromosomes found in each gamete. One complete chromosome could not be made from 23 genes.

6. **B** is the correct answer because in the human female body, fertilization normally takes place in the oviduct, or fallopian tube. In this tube the mature egg meets the mature sperm.

   A is not correct because the vagina is the tube where the sperm first enters the female body.

   C is not correct because the uterus is the muscular organ where the fertilized egg implants itself for development.

   D is not correct because the ovary is the female reproductive organ that produces mature eggs.

Answers Explained Section 7-10

1. **C** is the correct answer because the skin is the human body's first line of defense against disease. The skin provides a physical barrier against the entrance of pathogens.

   A is not correct because white blood cells are part of the defense system, but are not activated unless the pathogen gets by the skin and inside the body.

   B is not correct because hydrochloric acid present in the stomach kills many germs that have entered the body by way of the mouth.

   D is not correct because the lymphatic system contains many white blood cells that attack pathogens that have entered the body, especially the blood system.

2. **C** is correct because the neutrophil white blood cells are found in the highest percentage in the body. Neutrophil cells make up about 64% of all the white blood cells.

   A is not correct because eosinophil white blood cells make up about 5% of the total number of white blood cells in the body.

   B is not correct because monocyte white blood cells make up about 5% of the total number of white blood cells in the body.

   D is not correct because basophil white blood cells make up about 1% of the total number of white blood cells in the body.

3. **C** is the correct answer because the helper T cells initiate the immune response in the human body. The helper T cells are sometimes called the generals of the immune system.

A is not correct because killer T cells are activated by helper-T cells and are responsible for finding and destroying pathogens in the body.

B is not correct because B cells produce antibodies that are used to mark pathogens for destruction by other white blood cells. Some B cells also remember the pathogen that attacked the body, this speeds the future attack by the body on these now recognized pathogens.

D is not correct because antibodies are produced by the B-cells and in some cases destroy pathogens once they come into contact with the invading cell, but in many cases the antibodies just mark the pathogenic cell for destruction by other white blood cells.

4. **B** is the correct answer because the name of the substance that enters the body and triggers an immune response is called an antigen.

   A is not correct because antibodies are produced by the B cells and in some cases, destroy pathogens once they come into contact with the invading cell, but in many cases the antibodies just mark the pathogenic cell for destruction by other white blood cells.

   C is not correct because mucus is a thick, sticky, fluid produced in various parts of the body that traps germs for destruction by white blood cells.

   D is not correct because histamine is a chemical produced in the body that causes swelling of tissue from the influx of blood to a specific area.

5. **B** is the correct answer because raising the body temperature causes most pathogenic organisms to reproduce more slowly. The creation of a fever enables the body's immune system to "catch up" in cell number to the invading cells.

   A is not correct because the cerebellum does not initiate the fever response in the human body. The part of the brain responsible for raising the body temperature is called the anterior hypothalamus.

   C is not correct because the number of white blood cells will increase in most normal cases of an attack by a pathogenic organism.

6. **B** is the correct answer because mucus is a thick, sticky, fluid produced in various parts of the body that traps germs for destruction by white blood cells.

   A is not correct because, in some cases, mucus can prevent the leakage of bodily fluids from the body, but this does not really serve any type of defense to disease. In fact, in some cases, retaining certain infected fluids can be harmful.

   C is not correct because mucus does not have a temperature-regulating feature.

D is not correct because mucus can trap harmful pathogens, but white blood cells are needed to do the digesting of these cells.

Answers Explained Section 7-11

1. **B** is correct because pollen would not be considered a pathogen because, by definition, a pathogen causes disease or a toxic response in the infected organism. Pollen can cause the production of histamine in some individuals, but not disease.

   A, **C**, and **D** are not correct because some bacteria, viruses, and fungi are capable of causing disease in an organism so they are possible pathogens.

2. **A** is the correct answer because pasteurization is the boiling and cooling of a substance to kill possible pathogens. Louis Pasteur first devised this method.

   B is not correct because distillation is the method of separating substances in a mixture by evaporation of a liquid and the following condensation of its vapor.

   C is not correct because a neutralization reaction is a reaction of an acid with a base to produce salt and water.

   D is not correct because sublimation is the phase change of a substance from a solid directly to a vapor.

3. **B** is correct because Louis Pasteur is responsible for the practice of boiling surgical equipment before surgery to kill possible germs present on these utensils.

   A is not correct because the present-day classification system of using binomial nomenclature names to identify organisms was devised by Carolus Linnaeus.

   C is not correct because Edward Jenner started the practice of vaccinating people to prevent disease.

   D is not correct because Robert Koch devised the postulates for the specific scientific steps that need to be taken to prove a specific germ causes a specific disease.

4. **C** is the correct answer because a microorganism does not have to be unicellular to be tested by Koch's postulates.

   A, **B**, and **D** are not correct because the same microorganism must be found in all people or animals that have a particular disease, the microorganism must be isolated and grown in a lab setting, and when a particular disease-causing microorganism is grown in the lab and then injected into a healthy animal, it must cause the same disease are all parts of Koch's postulates.

5. **C** is correct because Edward Jenner was the first person to vaccinate people against disease.

   A is not correct because Robert Koch formulated ideas concerning the study of disease.

   B is not correct because Charles Darwin is famous for his ideas on evolution.

   D is not correct because William Harvey is noted for his work on blood circulation.

Answers Explained Chapter 7 Exam

Matching Column

1. G
2. A
3. E
4. H
5. B
6. C
7. D
8. F
9. C
10. B
11. E
12. A
13. D

Multiple Choice

14. **C** is correct because coronary arteries feed the heart tissue with oxygen. A heart attack is caused by the lack of oxygen to the heart muscle when these coronary arteries are blocked.

    A, **B**, and **D** are not correct because if veins in the leg, blood vessels in the brain, or blood vessels in the kidney were blocked, it would be a serious medical condition but not called a heart attack, because the heart tissue would not be directly affected.

15. **B** is correct because another name for white blood cells is leukocytes.

    A is not correct because erythrocyte is the medical name for red blood cells.

    C is not correct because fibrin is the chemical produced by the platelets in the blood that initiates the clotting response.

    D is not correct because lymph is the name of the fluid that leaks out of the blood vessels in the human body.

16. **B** is correct because the epiglottis is the flap of tissue that covers the trachea when food or drink is swallowed to prevent these substances from entering the windpipe.

    A is not correct because the tonsils are lymph tissue that protects the throat from infection.

C is not correct because the alveoli are the small sac-like structures found in the lungs where gas exchange between the environment and the blood takes place.

D is not correct because bronchioles are the small air tubes that branch off from the bronchi and eventually connect to the many alveoli in the lungs.

17. **A** is correct because the section of the brain that controls breathing and the beating of the heart is the medulla. The medulla is sometimes referred to as the "primitive brain."

    B is not correct because the cerebellum is the section of the brain responsible for balance and coordination of muscle movements.

    C is not correct because the cerebrum is the section of the brain responsible for thought, memory, creative thinking, and other higher-level brain functions.

18. **D** is correct because the pituitary gland secretes hormones that influence the actions of many other glands in the body.

    A is not correct because the thymus gland is involved in the immune activities of the body.

    B is not correct because the thyroid gland secretes hormones that control the metabolic rate of the body.

    C is not correct because the adrenal gland secretes the hormone adrenaline, which stimulates the rate of the breathing and the pumping action of the heart.

19. **C** is correct because testosterone is the male hormone produced in the testes and is responsible for secondary male sex characteristics.

    A is not correct because estrogen is a female hormones that controls the development of secondary female sex characteristics.

    B is not correct because progesterone is a female hormone that prepares the uterus for implantation of the fertilized egg.

    D is not correct because adrenaline is a hormone secreted by the adrenal gland and is responsible for control of breathing and heart rate at times of stress.

20. **B** is correct because the tympanum (or ear drum) is the thin membrane located at the end of the auditory canal.

    A is not correct because the retina is the light-sensitive section of the eye that consists of two structures called rods and cones.

    C is not correct because the cochlea is a structure found in the inner ear responsible for sound reception.

D is not correct because the stirrup is a structure that is found in the inner ear that helps transmits sound waves from the ear drum to the cochlea.

21. **B** is correct because ligaments connect bones to other bones in the human body.

    A is not correct because tendons connect muscle tissue to bones in the human body.

    C is not correct because cartilage is a strong, connective tissue that helps support many parts of the human body.

    D is not correct because muscles are tissue in the body that control internal and external movements of the organism.

Short Response

22. Sperm is created in the testes of the human male with help of the hormone testosterone. The sperm then leaves the testes and travels through two tubes called the vas deferens. The prostate and seminal glands secret fluids into the vas deferens tubes as the sperm pass through. The prostate gland secretes an alkaline solution to protect the sperm once they enter the acidic vagina, and the seminal gland secretes a nourishing solution to help give the sperm the energy needed for its trip to the egg. The two vas deferens tubes, which are now filled with this sperm and gland mixture, now called semen, join at the tube in the penis called the urethra. Semen leaves the body by way of this tube in the penis called the urethra.

23. The human immune response is the body's internal defense mechanism against disease. When pathogens have invaded the internal body, a chemical called interleukin-1 is released by macrophage white blood cells to alert another type of white blood cell called helper T cells. From this point, helper T cells control the immune response by sending chemical messengers to other white blood cells, initiating them to action. Killer T cells are one of these types of white blood cells activated by the helper T cells. Killer T cells look for and destroy foreign substances in the body. White blood cells called B cells are also activated at this point. B cells have two functions in the immune response. The first function is the production of antibodies, which mark pathogens for destruction by other white blood cells and, in some cases, these antibodies will actually destroy the pathogens themselves. The second function of some B cells is to remember the shape of the pathogen that attacked the body, this will help the body prepare quicker from an invasion by a similar pathogen in the future.

24. The following are the steps of Robert Koch's postulates:

    a. The same microorganism is found in all people or animals that have the disease.

b. This microorganism must be isolated and grown in a lab setting.

c. When the microorganism that is grown in the lab is injected into a healthy animal, it must cause the disease.

d. The same microorganism should be obtained from the new diseased animal.

These postulates have provided scientists with an organized approach for identifying the actual agents of disease. When these agents are identified people can then develop ways to combat them.

25. Human breathing is accomplished by the movement of a sheet of muscle in the lower abdomen called the diaphragm. The movement of the diaphragm causes the inhalation and exhalation of gases into the lungs. When someone inhales gases from the environment the diaphragm contracts, this flattens the muscle moving it lower in the chest cavity. A slight vacuum is created inside the chest area creating a space of lower pressure inside the chest compared to outside the body. Air floods into the lungs where there is lower pressure.

When someone exhales gases from the environment the diaphragm relaxes and this brings this muscle back to its normal resting position. The space in the chest cavity becomes smaller during this action causing the air pressure inside to increase. Now the air pressure is larger inside the chest than outside the body and the air rushes out of the lungs into the environment.

26. The medical condition called "heartburn" is really better named acid reflux. Sometimes acids in the stomach are pushed up the esophagus during digestion of food. These acids can irritate the lining of the esophagus, causing a burning sensation in the chest near the heart. It is not the heart that is burning, but the lining of the esophagus tube that is located in front of this structure in the body.

The structure called the cardiac sphincter is a muscle that controls the movement of materials into the stomach from the esophagus. When food enters the stomach from the esophagus, the sphincter muscle usually closes so that food and acid can not move up the esophagus. In some individuals this cardiac sphincter muscle becomes weak and this allows digested food and acid up the esophagus, which can cause "heartburn."

A Survey of Life

Scientists have classified about 2 million species of organisms on Earth. New organisms are being identified and classified every day. In fact, many scientists believe that there are probably more than 10 million species on the planet, which means we have classified only about 20% of the life that exists on this planet. Classifying and organizing these many different organisms into groups is important for biological study and retrieval of information.

Lesson 8-1: How Is Life Classified?

The particular branch of biology that concerns itself with grouping and naming organisms is called **taxonomy**, and the people who classify newly discovered organisms are called ***taxonomists***. Organisms are usually placed into specific groups based on comparative anatomy, DNA, chemical makeup, and their geographic locations. These factors are used to show if a group of organisms have an evolutionary relationship.

Classification and groupings make the study of life easier. For example, if you went into a supermarket and had to purchase a loaf of bread, a gallon of milk, and a box of dry cereal. Even though most supermarkets contain thousands of different types of food, you could quickly go into the store and find what you need to buy. The reason for this is that all the food in the store is grouped; dairy products are found in one location, dry cereal is found another location, and breads are found in yet another. When a scientist needs to study a particular organism from the group of the more than 2 million identified, a system for arrangement becomes very helpful.

In the modern classification system all organisms are given a specific name called **binomial nomenclature**. Scientists around the world, regardless of the language spoken, know this naming system. This naming system uses two Latin words to describe the organism in question. The first word, which is always capitalized, is the name of the ***genus*** the organism is grouped into. The second word, not capitalized, is the name of the ***species***. These two names that classify the organism in question are always italicized when printed. Latin is the language used by all scientists because it is not spoken in conversation and, therefore, does not change.

Example

The scientific name of the white oak is *Quercus alba*.

Quercus is the genus grouping of this organism which means "oak."

Alba is the species name of this organism and means "white."

Another organism in the same genus is *Quercus rubra*, which is the scientific name of the red oak tree.

••••••

By giving all organisms a binomial nomenclature name, confusion brought about by common names is avoided. For instance, the organism ***Panthera concolor*** is known by many common names such as mountain lion, catamount, and cougar. Two people could be talking about the same organism and not even know it if they were using different common names for the same species.

Binomial nomenclature was started in the late 18th century by a Swedish botanist by the name of Carolus Linnaeus (1707–1778).

Binomial Nomenclature of Various Organisms

Common Name	Genus	Species
human being	*Homo*	*sapien*
house cat	*Felis*	*domesticus*
domesticated dog	*Canis*	*familiaris*
house sparrow	*Passer*	*domesticus*
earthworm	*Limbricus*	*terrestris*

Table 8.1

A group of organisms is called a ***taxon*** (plural, taxa). The taxa created are arranged from very broad groups to very specific groups. The broadest taxon is called the ***kingdom***. All the species in a kingdom have some common traits, but some species within the taxon will have more characteristics in common then other members within the group. The next broadest taxon is the ***phylum***, which contains less total species than a kingdom, but a greater percentage of these organisms have more common characteristics. This grouping of taxons continues from phylums to ***classes***, ***orders***, ***families***, ***genera***, and ***species***. A ***class*** is a taxon of similar ***orders***, while an order is a taxon of similar ***families***, and so on. Every organism classified so far on this planet is not only given a binomial nomenclature name, but is also classified in the seven taxons already mentioned. Table 8.2 on page 271 shows the seven taxons of some organisms.

Common Name	Human	House Cat	Gray Wolf	Chimpanzee
Kingdom	animal	animal	animal	animal
Phylum	Chordata	Chordata	Chordata	Chordata
Class	Mammal	Mammal	Mammal	Mammal
Order	Primate	Carnivore	Carnivore	Primate
Family	Hominidas	Felidae	Canidae	Pongidae
Genus	*Homo*	*Felis*	*Canis*	*Pan*
Species	*sapiens*	*domesticus*	*lupus*	*troglodytes*

Common Name	Fruit Fly	Corn	Mushroom
Kingdom	animal	plant	fungi
Phylum	Arthropoda	Anthophyta	Basidiomycota
Class	Insect	Monocotyledores	Hymer
Order	Diptera	Commelinales	Agaricales
Family	Drosophilidae	Poaceae	Agaricaceae
Genus	*Drosophila*	*Zea*	*Agaricus*
Species	*melanogaster*	*mays*	*campestris*

Table 8.2

You can see in Table 8.2 that the last two taxons an organism is placed into is its binomial nomenclature designation.

The seven taxon system of classification had originally been designed to divide all of life into five manmade groups called kingdoms. You can see in Table 8.2 that the kingdom is the first and broadest group an organism is placed within this system. Life is divided into the five kingdoms based on the criteria shown in Table 8.3 on page 272.

These five kingdoms have been a standard in taxonomy since the late 1960s, when an ecologist by the name of Robert Whittaker established them. In recent years with the advances in biology come important changes that are being considered for this grouping method.

Based on Whittaker's system, the monera kingdom contains all the bacteria found on the planet. In the 1970s, tests were done on organisms in the kingdom monera. These tests showed that all bacteria within the group were not as similar as had been originally thought. It was discovered that one group in particular, called the ***archaebacteria*** (ancient bacteria), after studying the genome, is very different from most other bacteria. This group of bacterial cells was so different that a new kingdom called the ***archaebacteria*** kingdom was established for them.

Kingdom Name	Traits of Organisms Placed in This Kingdom	Examples of Organisms Placed in This Kingdom
Animal	The cellular makeup is eukaryotic. All organisms are multicellular, ingestive heterotrophs.	humans, insects, birds, dogs, cats
Plant	The cellular makeup is eukaryotic, all are multicellular, photosynthesizing organisms.	trees, bushes, herbs, grasses
Fungi	The cellular makeup is eukaryotic. Most are multicellular, absorptive heterotrophs.	mushrooms, mold, mildew
Monera (Bacteria)	The cellular makeup is prokaryotic. Some organisms are autotrophic while others are heterotrophic.	E.coli and streptococcus
Protist	The cellular makeup is eukaryotic. Some organisms are autotrophic while others are heterotrophic.	ameba, paramecium, euglena, algae

Table 8.3

Another problem of the five-kingdom system was that the protista kingdom really was not a very natural grouping of organisms. Many of the organisms placed into this group were done so because they did not fit into any other kingdom, not because they had many similarities to the other organisms within their kingdom. In a very recent update, Grenbank, a federal agency that collects genetic information about life on this planet, devised a three-superkingdom system.

Grenbank's Superkingdom System

Kingdom	Characteristics
Archaea	All the ancient bacterial cells that mostly live in extreme environments such as near deep-ocean hydrothermal vents or in very high saltwater content ponds.
Eubacteria	Bacterial cells that are less complex than the archaebacteria.
Eukaryota	This kingdom contains all the organisms of the animal, plant, fungi, and protest groupings. They are all placed into this group because they are made up of similar organelle-filled cells. The organisms in this kingdom are not further divided in this system because of the number of differences concerning their origins that has been uncovered.

Table 8.4

It is important to understand that even this new three-superkingdom system is not fully supported by the scientific community. As you can see, the naming and classifying of life on this planet is a constantly changing endeavor based on the recent finds in science.

Practice Section 8-1

1. What is a close estimate of the number of classified organisms on the planet?
   a) 100,000 b) 500,000 c) 2 million d) 20 million

2. According to binomial nomenclature, which two taxons are used to represent the organism?
   a) kingdom and phylum c) genus and species
   b) class and order d) family and genus

3. What binomial nomenclature name for human beings is written correctly?
   a) *Homo Sapien* b) *homo Sapien* c) Homo sapien d) Homo sapien

4. What is the correct order for the taxons of life from most general to most specific?
   a) kingdom, phylum, order, class, family, species, genus
   b) phylum, kingdom, class, order, genus, family, species
   c) kingdom, phylum, class, order, family, genus, species
   d) class, phylum, order, kingdom, family, genus, species

5. According to the five-kingdom system, what kingdom contains organisms whose structure is composed of *prokaryotic* cells?
   a) fungi b) animal c) bacteria d) plant e) protista

6. What two organisms are the most closely related?
   a) *Pan troglodyte* and *Zea mays*
   b) *Zea mays* and *Canis lupus*
   c) *Canis lupus* and *Felis domestica*
   d) *Felis domestica* and *Pan troglodyte*
   e) *Canis familiaris* and *Canis lupus*

7. What are the differences between Archaebacteria and Eubacteria?

8. What are some examples of human diseases caused by bacteria?

9. According to their body structure, how are bacteria classified?

Lesson 8-2: The Animal Kingdom

All animals are multicellular, eukaryotic, heterotrophic organisms. All members of the animal kingdom also produce a protein, found, for example, in human skin, called ***collagen***. Other organisms outside of the animal kingdom do not seem to ever make collagen. All animals also develop from a ***blastula*** cell stage (except sponges). Most members of the animal kingdom possess nervous systems, while no other organism found outside of this kingdom has one. In the seven-taxon system of classification, the next broadest grouping after kingdom is phylum. There are 32 phylums under the animal kingdom. Next, we will look at some of the members of some of these phylums.

Phylum Porifera

The dominant members of this phylum are the sea sponges. All members of this phylum have asymmetrical body shapes and lack complex tissues and organs. The body walls of organisms in this phylum are created of two cell layers that have many small holes in them. The inside of the body cavity, which is usually cup-shaped, is lined with cells called ***chaonocytes.*** The choanocytes capture food from the water that cycles into and out of the body cavity from the environment. There are more than 8,000 species in the phylum Porifera.

Phylum Cnidaria

Jellyfish, sea anemones, corals, and hydras are some of the representatives of this phylum. All members of this phylum have radially symmetrical body plans that are created of two cell layers. Between these two cell layers is a soft layer that can be very thin or very thick depending on the species in question.

The members of this phylum have what is sometimes called a "two-way" digestive tract. This means that the opening from the external environment to the inside body cavity acts as both a mouth, where food enters, and as an anus, where waste leaves. All cnidarians have structures called ***nematocysts.*** Nematocysts are stinging cells used to paralyze their prey. There are more than 10,000 species in the phylum Cnidaria.

Bilateral Symmetry

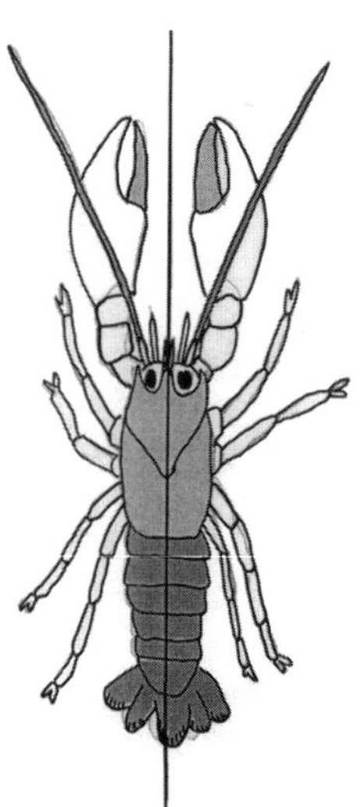

Phylum Annelida

This phylum contains the segmented worms. All members of this group have bilateral symmetry and a ***coelom*** (a fluid-filled cavity arising in the mesoderm, which separates the body wall from the gut wall, allowing independent movement of each). All members of this phylum have well-developed circulatory, excretory, and nervous systems. There are more than 15,000 species in this phylum.

Phylum Mollusca

Squids, octopuses, mussels, slugs, snails, clams, and oysters are some representatives of this phylum. Members of this group have common body parts, such as the molluscan foot, a mantle, the feeding structure called the ***radula***, the breathing apparatus called the ***ctenidia,*** and the gathered internal organs called the ***visceral***

mass. Most mollusks are identified by their calcium-based shell (snail) or in the case of bivalves, (clams) two shells. There are a great deal more than 100,000 species in this phylum.

Phylum Arthropoda

Some members of this phylum include insects, spiders, lobsters, crabs, shrimp, and scorpions. This phylum is home to more than 90% of all the identified animals on this planet. All members of this group have segmented bodies with paired, jointed appendages ("arthro" = jointed and "poda" = leg). They are all bilaterally symmetrical and have external skeletons called **exoskeletons** that are made of a protein called chitin. The largest taxon within this group is the class Insecta. In this taxon class alone there are more than 750,000 species identified.

Phylum Echinodermata

Some members of this phylum include starfish, sea urchins, sand dollars, and sea cucumbers. The adult members of this phylum are radially symmetrical, while the immature stages of most are bilaterally symmetrical. Echinodermata means "spiny-skinned" and members of this group are just that, covered in spines. All members have what is called ***pentaradial symmetry***, which means they have five radiating parts from a center structure. This phylum contains more than 6,000 species.

Phylum Chordata

Sample of Taxon Classes in Phylum Chordata

Class	Description	Examples
Amphibia	Almost all members go through a water-dwelling and a land-dwelling stage. These organisms have no scales and almost no skeleton. Their skin is moist and can be used for the absorption of oxygen. Fertilization is external.	toads, frogs, salamanders, and newts
Reptilia	All members have dry skin that is covered in scales. Most members lay shell-covered eggs. A well-developed skeleton structure. Fertilization is internal.	extinct dinosaurs, lizards, turtles, snakes, crocodiles, and alligators
Aves	All members of this class have bodies that are covered in feathers, forelimbs that are wings, are endothermic, and lay shelled eggs.	turkeys, owls, pheasants, eagles, ducks, penguins, and robins
Mammals	Hair found on parts of the body, offspring are fed milk produced from mammary glands.	humans, squirrels, horses, gorillas, whales, and elephants

Table 8.5

Representatives of this phylum include all fish, reptiles, amphibians, birds, and mammals. The members in this group get their name from the structure called the ***notochord***. The notochord is a long, stiff rod of cartilage that forms beneath the nerve cord in the dorsal area. In most members of this phylum the notochord is replaced in early development by a hollow backbone called the ***vertebral column***. The vertebral column supports and protects the nerve cord that runs along the back of the organism. The dorsal nerve chord is hollow and sends nerves throughout the body of the organism. Most members of this phylum have a somewhat developed brain surrounded by a bone case called the ***skull***. There are more than 42,000 species in this phylum. Some of the taxon classes found under the Chordata phylum are explained in detail in Table 8.5 on page 275.

Practice Section 8-2

1. Which of the following characteristics is not descriptive of an organism in the animal kingdom?
   a) Organisms in this kingdom have structures composed of eukaryotic cells
   b) All members of this kingdom produce a protein called collagen.
   c) All members of this kingdom are heterotrophic.
   d) All members of this kingdom are unicellular.
2. Which of the following organism is not classified in the animal kingdom?
   a) human
   b) earthworm
   c) slug
   d) jellyfish
   e) all are members of the animal kingdom
3. What phylum found in the animal kingdom contains organisms that have the following characteristics: bilateral symmetry, a coelom, and well-developed circulatory, excretory, and nervous systems.
   a) Cnidaria b) Mollusca c) Annelida d) Arthropoda
4. What phylum in the animal kingdom are humans classified into?
   a) Mollusca b) Annelida c) Chordata d) Echinodermata
5. Members of the phylum Chordata are distinguished by
   a) an exoskeleton c) nematocysts
   b) pentaradial symmetry d) vertebral column
6. Which of the following classes is not found in the phylum chordata?
   a) amphibian b) reptile c) bird d) mammal e) all of them

Lesson 8-3: The Plant Kingdom

All plants are multicellular, autotrophic organisms. All members of the plant kingdom are made of eukaryotic cells that contain ***cellulose*** Some plant cells also contain a green pigment called ***chlorophyll***. During their lives, diploid animals produce haploid gametes. Their gametes will, in some cases, fuse to produce diploid animals once again. In plants two different forms of a plant complete this process during the generation. Nearly all plants experience what is called an **alternation of generations**. Plants alternate between a haploid and a diploid stage. The haploid plant usually produces the gametes (gametophyte), and the diploid plant produces the spores (sporophyte). The gametes join to produce the zygote, which then develops into the sporophyte plant. The spores produced by the sporophyte plant germinate and produce the gametophyte plant, completing the cycle. In plants that produce flowers, the embryo sac and the pollen represent the female and male gametophyte generations, respectively.

In the seven taxon system of classification, the next broadest grouping after kingdom for plants is called ***division***. The division is similar to the phylum taxon for the other kingdoms. There are more than 10 divisions under the plant kingdom. We will look at some of the members of some of these divisions next.

Division Bryophyta

The members of this division are small, nonvascular plants that live in moist habitats. The haploid gametophyte generation is the dominant form. Mosses and liverworts make up the more than 15,000 species in this division.

Division Pterophyta

The members of this division all contain vascular systems. The sporophyte generation is dominant. The representatives of this division do not produce seeds but contain spores on their leaves. There are about 12,000 species of fern in this division.

Division Coniferophyta

The members of this division all contain vascular systems. The leaves of these organisms are needle-shaped. Naked seeds are produced in cones that develop on the plant. The sporophyte generation is dominant. Spruces, firs, and yews make up the more than 500 species of this division.

Division Anthophyta

This division by far has become the most widespread and dominant in the plant kingdom. All members are what are called ***angiosperms***. Angiosperms are plants that produce flowers as reproductive structures. Ovaries enclose ovules, and the fertilization of the ovule produces the seed, while the surrounding ovary becomes the fruit. The sporophyte generation is dominant. There are more than 240,000 species in this division and include such plants as orchids, maple trees, oak trees, roses, tulips, sunflowers, and water lilies.

Practice Section 8-3

1. What statement below is not true about the plant kingdom?
   a) All members of this kingdom are autotrophic.
   b) All members of the kingdom have cells that contain cellulose.
   c) All members contain reproductive structures called flowers.
   d) All members of this kingdom are multicellular.
2. Ferns are members of what division in the plant kingdom?
   a) Anthophyta b) Bryophyta c) Pterophyta d) Coniferophyta
3. What division in the plant kingdom contains the most classified members?
   a) Bryophyta b) Pterophyta c) Coniferophyta d) Anthophyta
4. Alternation of generation in plants means that
   a) the plant alternates between autotrophic and heterotrophic nutrition
   b) the plant reproduces asexually
   c) the plant alternates between a haploid and diploid stage
   d) all of these
5. What is the name of the green pigment used by members of the plant kingdom to capture energy from sunlight?
   a) chloroplast b) chitin c) cellulose d) chlorophyll
6. Plants that are classified as angiosperms contain
   a) flowers b) pinecones c) spores d) "naked" seeds

Lesson 8-4: Plant Structure

Some of the most common plant species found on Earth today are found in the division Anthophyta. In this taxon division, the plant species produces seeds for reproduction and contain well-developed transport systems for moving food and water within their structures. In this lesson, we will look more closely at the tissue structures of this particular plant division.

Plant Cells

Like all living organisms on this planet, plants are composed of cells. The cells of a plant contain many of the same organelles as the cells of an animal. Table 8.6 shows some of the differences.

Comparison of Plant and Animal Cells

Structure	Plant Cell	Animal Cell
Cell wall	Found surrounding and supporting the cell membranes of plant cells.	Cell walls are not found in animal cells.
Chloroplasts	Organelles that contain the photosynthesizing chemical chlorophyll.	Chloroplasts are not found in animal cells.
Vacuoles	Vacuoles are very large inside the cells of plants.	Vacuoles are very small inside the cells of animals.
Centrioles	Centrioles are not found in plant cells.	Centrioles are involved in animal cell reproduction.

Table 8.6

The structure of a seed plant is usually divided into three principal organs: the root, the stem, and the leaf. Each of these organ systems is then composed of groups of cells with distinct structures and functions.

In most plants there are three types of cells. Table 8.7 gives a brief comparison of these cells.

Three Types of Plant Cells

Cell Type	Structure	Function
Parenchyma	These cells are spherical in shape and have flexible cell walls. The vacuoles in these cells are usually very large. These cells are usually the most abundant cells in the plant.	The two main functions of these cells are photosynthesis and food storage.
Collenchyma	These cells have cell walls that are irregularly thickened with cellulose and pectin.	These types of cells are usually located right beneath the epidermis and function as a support mechanism for the plant.
Sclerenchyma	The cell walls of these cells are very thick and rigid. Some of these types of cells are long and thin while others are small and circular.	These types of cells are the main supporting tissue of plants and also form much of the vascular system with many plants.

Table 8.7

The three types of cells previously mentioned are then arranged in groups or systems within the structure of the plant.

The three tissue systems of cells in the plant are: dermal tissue, vascular tissue, and ground tissue.

- **Dermal tissue: The "skin" of the plant**

  Most plants are covered in a single layer of cells called ***epidermal cells***. These epidermal cells are a type of parenchyma cell. The epidermal cells make up the epidermis of the plant. The epidermis is like the skin of the plant. This collection of cells provides a protective coating on the plant. Some of these epidermal cells produce a ***waxy cuticle*** that helps the plant prevent the loss of water through evaporation. Some plants have structures called ***trichomes*** extending from the epidermal cells on the stems and leaves. These extensions of the epidermal cells give the area where they are present a fuzzy appearance. Studies have shown that these hairlike extensions reflect sunlight, lower the temperature of the leaves, and lower the rate of water loss. In some plants the trichomes are defense mechanisms against insect predators.

- **Vascular tissue: The transports system of the plant**

  The vascular tissue transports all the food, water, and minerals throughout the plant. The vascular tissue is composed of three different types of sclerenchyma cells and also parenchyma cells. **Xylem** and **phloem** are the two major types of vascular tissue. Xylem is tube shaped plant tissue that transports mineral and water from the roots of the plant to the highest leaves and stems. Phloem is plant tissue composed of tube cells that transport the food produced by the leaves to the rest of the plant structure.

- **Ground tissue: Food and support**

  The ground tissue consists of all the other cells of the plant that are not involved in the vascular system, or epidermis. Ground tissue consists of parenchyma, sclerenchyma, and collenchyma cells. The cells of this tissue type are involved in photosynthesis, food storage, and supporting the structure of the plant.

Parts of the Plant

The leaf

Leaves are the structures where most photosynthesis occurs in plants. The upper surface of most leaves captures the energy from the sun and the bottom of these same leaves are dotted with thousands of holes that collect carbon dioxide and release oxygen and water. Most leaves contain a variety of cells that enable plants to most efficiently complete the process of photosynthesis. The diagram on page 281 shows a cross section of a typical leaf.

The ***epidermis*** comprises the outer layer of cells on the surfaces of the leaf. These cells are bricklike in appearance and are usually covered in a waxy layer called a ***cuticle***. The epidermal cells are clear and allow light to penetrate into the interior cell regions of the leaf. The epidermal cells, located on the bottom of the leaves, have many small openings called **stomata**. These openings in the bottom of the leaf allow the plant to take in carbon dioxide gas from the environment for photosynthesis, and also allow the plant to excrete oxygen gas and excess water vapor. The process by which a plant gives off water vapor through the stomata into the environment is called **transpiration**. On very warm days significant amounts of water can be lost through the stomata on the leaves. There are two cells found on either side of each stoma. These cells are called **guard cells**, and they regulate the opening and closing of these holes. Sometimes these guard cells will close the stomata to conserve the loss of water during the warm daylight hours, and generally the stomata will close at night because the plant does not need environmental gases for photosynthesis at this time.

Cross Section of a Leaf

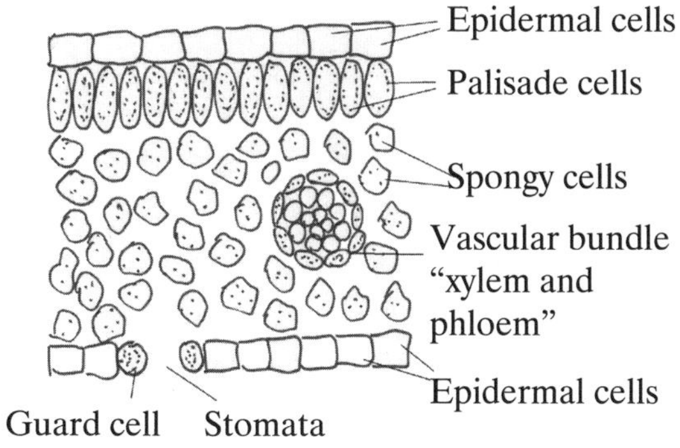

The tissue layer found between the epidermal cell layers is called the ***mesophyll***. The mesophyll consists of two different cell types: palisade cells and spongy cells.

Palisade cells are located in columns under the upper epidermal layer of the leaf. These cells are loaded with chloroplasts that are in turn full of chlorophyll, and are the location where most photosynthesis occurs in the leaf.

The spongy cell layer is a collection of irregular shaped cells loosely arranged below the palisade cells. These spongy cells also contain chloroplasts for photosynthesis, but their major function is the collection of water and gases from the environment for the process of photosynthesis.

The stem

The stem is the structure that supports the leaves and flowers of the plant and contains vascular tissue that will move food, water, and minerals to their proper locations within the plant. The stem can range in size and strength from the thin, soft, green, stem of a dandelion to the thick, hard, brown, wooden stem of an oak tree. The soft green stems on plants are called ***herbaceous stems***, while the hard thick stems of trees are called ***woody stems***.

Many species of plants also use the stem as an area for food storage. The underground stem of the potato plant is called a "tuber." The tuber is a food storage site that humans make use of when harvesting this plant.

The root

The root is the part of the plant that grows underground to anchor the plant in the soil and to collect water and minerals. There are two major types of roots: **fibrous** and **taproots**. In plants that have taproots there is usually one thick primary root that grows downward from the plant. In plants that have fibrous roots, all the roots are usually about the same size and are much thinner and grow in all directions from the base of the plant.

Types of Roots

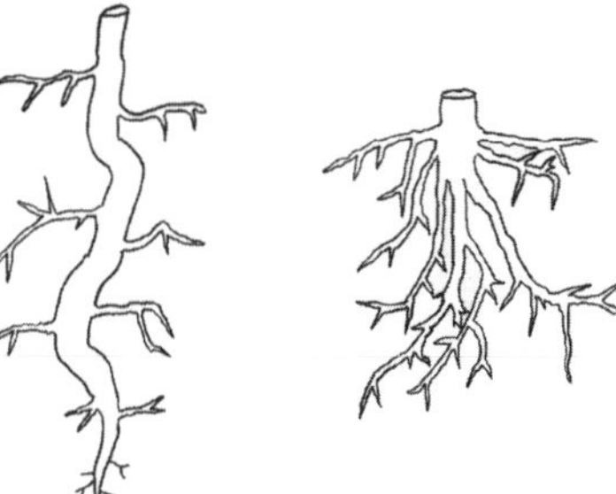

The root of a plant has many functions. Roots anchor the plant to the soil so that the leaves and flowers will be supported in the air. Roots absorb water and minerals from the soil. Some plants will store large quantities of food in their root structures.

The structure of the root itself is divided into regions that contain distinct cells that have specific functions for the plant.

Starting at the base of the root is the **root cap**. The root cap is a collection of cells in the shape of a thimble that grows on the tips of the roots. As the root grows, the root cap pushes its way through the soil. Many cells on the root tip are destroyed in this process and their remains begin to form the root cap. As the root grows it also undergoes the process of respiration. Carbon dioxide gas is given off by the root tips during this process and forms a weak acid with water in the soil called carbonic acid. This carbonic acid helps weaken the soil surrounding the root tip and also aids in the absorption of water into the root itself.

Parts of a Plant Root

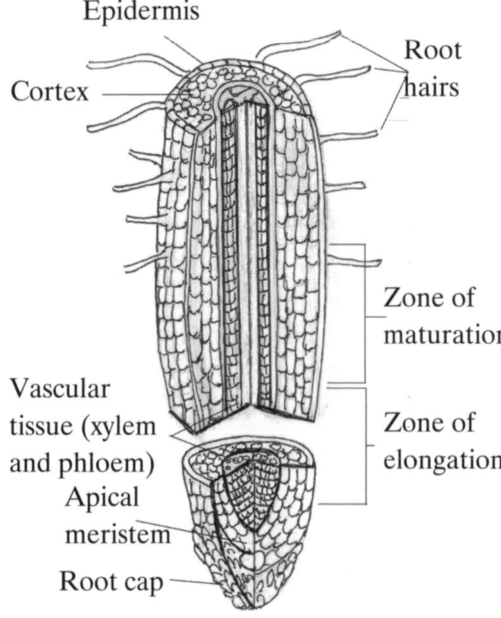

The surface of the root is covered in epidermal cells. Many of these epidermal cells have hairlike extensions called **root hairs**. Root hairs help the plant absorb water and minerals by providing increased surface area to the root. The water and minerals that enter the root hairs are channeled to the xylem tubes, which bring the water to the other parts of the plant.

The inner cells of the plant root are called the **cortex**, which consists of parenchyma cells. The cortex allows for movement of water, minerals, and gases from the epidermis of the plant root to the inner regions of the plant. The cortex also is a storage site for food produced in the plant. A good example of the storage ability of the cortex region is the carrot. The carrot is the enlarged cortex region of the taproot of a carrot plant.

Practice Section 8-4

1. Plant cells and animal cells have many of the same organelles inside. Which of the following organelles is only found in a plant cell?

   a) vacuole b) centriole c) chloroplast d) cell membrane

2. Plants contain three types of cells. Which cell type is most responsible for food production and storage?

   a) collenchyma c) sclerenchyma

   b) parenchyma d) prokaryotic

3. Plants cells are constructed into three tissue types. Which tissue type has the function of transport within the plant?

   a) dermal tissue b) ground tissue c) vascular tissue d) epidermal

4. What is the collection of "tube" cells that transport food through all vascular plants?

   a) xylem b) epidermal c) phloem d) trichomes

5. What is the name of the small openings at the bottom of plant leaves where most transpiration occurs?

   a) cuticle b) guard cells c) stomates d) trichomes

6. What structure found on plants increases the surface area of the root?

   a) stomates b) root hairs c) guard cells d) mesophyll

Lesson 8-5: Plant Reproduction

In the plant kingdom, the most common type of reproduction involves the use of flowers. Flowering plants are part of the division Anthophyta, which has more than 250,000 of the 270,000 species in the kingdom. We will now look more closely at reproduction within this kingdom and how it occurs.

The reproductive organ of plants in the Anthophyta division is the flower. The flower contains the male and female sex cells, or gametes, for reproduction. To better understand how sexual reproduction occurs in these types of plants, let us look at the various parts of this reproductive organ—the flower.

The **sepals** are the leaves that protect the flower when it is a bud, before the flower has opened. All the sepals together are called the ***calyx***.

Parts of the Flower

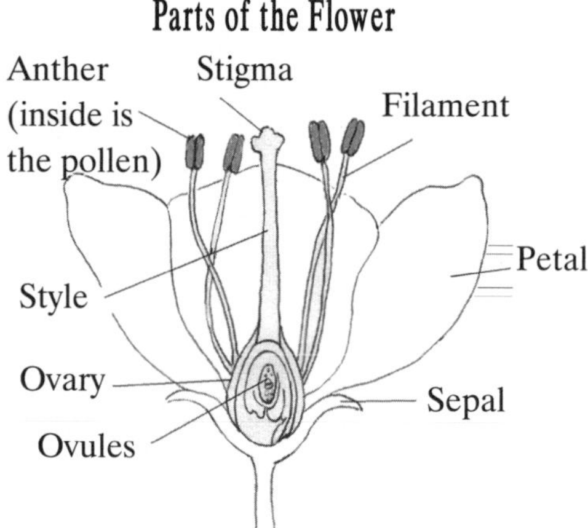

The **petals** are the next circles of flower leaves. All the petals make up what is called the ***corolla***. The petals of many flowers are very colorful and smell good, to attract birds and insects for pollination.

The ***stamens*** are located inside of the sepals and petals of the flower. The stamen are the male parts of the flower and each is made up of the filament, anther, and inside the anther the developing pollen or male sex cells.

The ***pistil*** usually lies in the center of the stamen. The pistil is the female part of the flower. The **stigma** is the structure of the pistil that collects the pollen. Pollination is the name of the process where pollen is transferred to the stigma. The ***style*** is the long, thin tube that holds the stigma at the top of the pistil to better collect the pollen. At the base of the style is the **ovary** that houses the **ovules**, or female sex cells. If the ovule is fertilized by pollen and becomes a zygote, the ovule will develop into the seed of the plant and the ovary surrounding it becomes the fruit of the plant.

Development of the Pollen and Ovules

All the cells of a plant have a diploid (2n) number of chromosomes. This means they contain the full number of chromosomes for the particular species. As the pollen grains develop inside the anther of the stamen, they undergo the process of meiosis. This eventually leads to each pollen grain having a nucleus with a monoploid (n) number of chromosomes.

The process of meiosis is also occurring inside of the ovule of the flower. A structure called the embryo sac develops within each ovule, and inside of this sac three monoploid structures form: the egg nucleus and two polar nuclei.

Early Development of the Seed

Inside the ovule of the ovary, an egg nucleus and two polar nuclei are held inside an embryo sac. When the pollen sperm fertilizes the egg, it begins to divide by the process of mitosis. This cellular division will eventually form the embryo of the plant in the ovule. A second pollen sperm that has entered the ovule will unite with the two polar nuclei to form the ***endosperm***. The endosperm becomes the food

Fertilization in the Plant Embryo

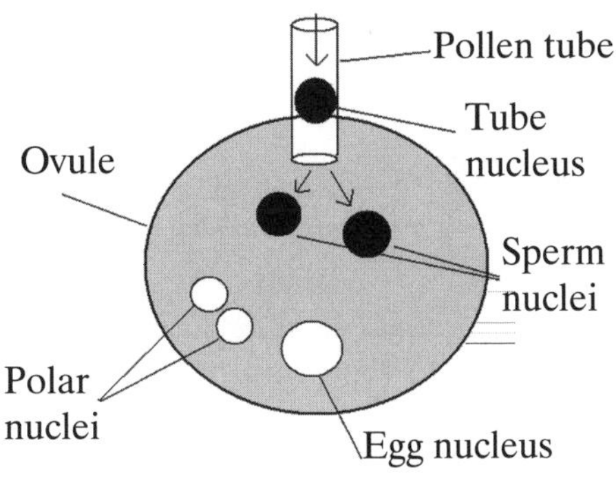

for the developing plant embryo. In many plants when the cotyledons form they will eventually take over the function of food storage from the endosperm.

Seeds

All seed plants contain a structure called an **ovule**. A seed is a structure formed by the development of the ovule following fertilization. Fertilization in seed plants occurs when the male sex cell, the pollen, enters the female sex cell, the ovule, and forms the zygote. The diploid zygote develops into the plant embryo. The **cotyledon** is the first leaf of the embryo of the seed plant. Cotyledons play an important function in the early stages of seed development.

These structures always act as the early food storage centers before the plant is making its own food through the process of photosynthesis. In many plants, the cotyledons also eventually become the first photosynthesizing parts of the plant. In the division Anthophyta (angiosperms), there are two classes called ***Monocotyledonae*** (monocots) and ***Dicotyledonae*** (dicots). Monocots are so named because they develop with one cotyledon, or seed leaf, and dicots are named because they develop with two cotyledons, or seed leaves. Some examples of monocot plants are grasses, palms, and orchids. Some examples of dicots include maples, elms, oaks, and roses. Table 8.8 on page 286 shows some of the characteristics of the two classes.

Fertilization in the Plant Embryo

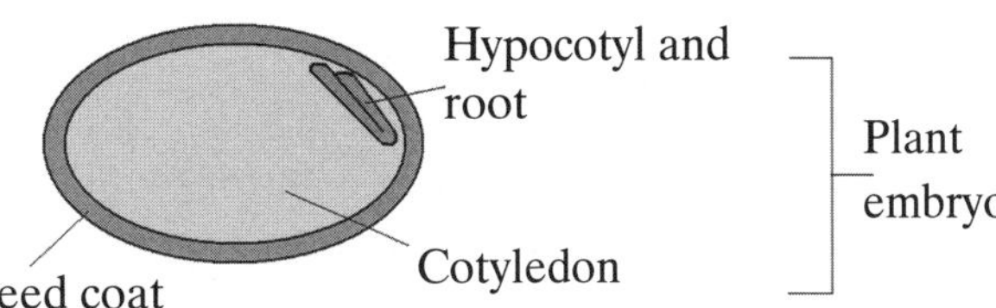

Germination of the Seed

Germination is when the embryo within the seed begins to develop into a new plant. For this to occur in most cases, the presence of water, the correct temperature, and other environmental factors must be correct for the particular plant seed species. In many species of plants, once the seed coat has softened, the embryo of the plant will start to emerge. Usually, the first part to appear is the embryo root, or ***radicle***. This structure will grow into the soil and eventually become the roots of the plant. The ***hypocotyl*** is the section of the plant stem above the radicle. In many plants, the seed leaves (cotyledons) will be attached to this growing hypocotyl stem. The ***epicotyl*** is the stem structure above the cotyledons of the developing plant.

Characteristics of Monocots and Dicots

	Monocots	**Dicots**
Leaves	Parallel veins within the leaves	Branched veins within the leaves
Flowers	Flower parts (petals, sepals, stamens, or pistils) usually in multiples of three	Flower parts (petals, sepals, stamens, or pistils) usually in multiples of four or five
Stems	Vascular tissue scattered throughout the inside of the stem	Vascular tissue arranged in a ring inside the stem
Seed leaves	Contains a single seed leaf or cotyledon	Contains two seed leaves or cotyledons
Roots	Root structures are fibrous	Root structure is a taproot

Table 8.8

Practice Section 8-5

1. There are many different species of plants in the plant kingdom, and not all of these plants reproduce in the same fashion. What method of reproduction is the most common in plants?
   a) spores b) pinecones c) flowers d) binary fission
2. Which of the following structures is a female part of the flower?
   a) sepal b) anther c) filament d) ovule
3. Which statement is true concerning the structure of pollen?
   a) Pollen contains the diploid number of chromosomes.
   b) Pollen is produced from the female part of the flower.
   c) Pollen is created in the anther of the flower.
   d) Pollen produces the structures known as ovules.
4. What part of a seed plant becomes the fruit?
   a) pollen b) ovule c) ovary d) sepals

5. The plant division *Anthophyta* contains two classes known as *Monocotyledonae* and *Dicotyledonae*. Which characteristic is only true for members of the *Monocotyledonae* class?

   a) Contain two seeds leaves.

   b) Flower parts are in multiples of fours and fives.

   c) Parallel veins within the leaves.

   d) Vascular tissue is arranged in rings inside the stem.

6. Which organism would be found in the class *Dicotyledonae*?

   a) palm tree b) lawn grass c) oak tree d) orchid

Lesson 8-6: Plant Growth

Plant Growth

Probably the most apparent response or visible movement concerning plants is that they grow, and grow and grow. In fact, unlike animals, plants continue to grow for their entire lives. The secret to this seemingly unlimited growth is the presence of what is known as **meristems**. A meristem is a specific region inside the plant where there are actively dividing cells. In fact, meristematic tissue contains the only plant cells that produce new cells by mitosis. The meristem regions of the plant contain undifferentiated cells, meaning these cells can, in essence, become any of the three types of tissue in the structure of the plant. The two major areas of the plant that you will find meristematic cells are the shoots and roots of the plant called ***apical meristems***, and the ***lateral meristems*** found in the vascular tissue in the stems of plants.

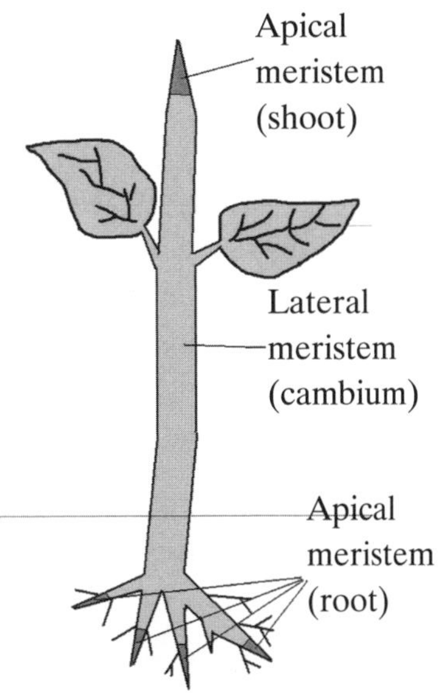

Growth of Shoots

The stems, or shoots, of the plant have a number of functions. The two major functions of the stems involve the movement of materials to various parts of the plant and to be supportive structures for the leaves and flowers of the plant.

As mentioned earlier, there are three types of plant tissue: epidermal, vascular, and ground tissue. Stems are made up of all three of these tissues.

The cells surrounding the outer portion of the stem are the epidermal cells. These dermal cells on the stem, just like the ones in the leaf, secrete a waxy cuticle for protection and water conservation.

The arrangement of tissue inside the epidermal cells then depends on the species of plant in question. **Monocot** and **dicot** plants have a different arrangement of ground and vascular tissue.

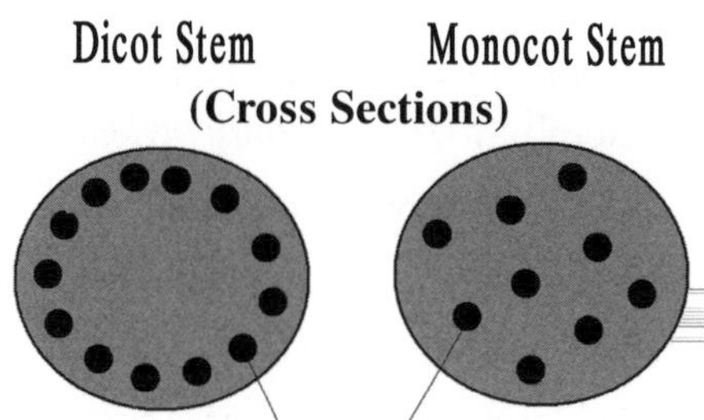

The inside of monocot stems is a uniform arrangement of parenchyma cells with a series of vascular bundles (xylem and phloem) scattered throughout the ground tissue of parenchyma cells.

Inside of a dicot stem, the ground tissue of parenchyma cells are located inside and outside of a ring of vascular bundles. The vascular bundles ring close to the inner epidermal cell layer. Between the vascular bundles and the dermal layer the parenchyma cells are called ***cortex***, while the parenchyma cells located inside the vascular bundles to the center of the stem are called ***pith***.

For the entire life of a plant, growth occurs at the tip of the shoots in the apical meristem cells. This type of growth is called ***primary growth***. This growth makes the plant taller and longer.

The stems of plants not only continually grow at their tips but also in an area called the ***lateral meristem tissue***. This thickening of the stem occurs in all dicot plants and some monocot plants in this division. This type of stem-widening growth is called ***secondary growth***. The lateral meristematic tissue in the stem is of two types: ***vascular cambium*** and **cork cambium**. The vascular cambium is located in between the phloem and xylem tissue. This cambium is the area where continually dividing cells produce more vascular tissue and, in turn, slowly thicken the stem. More xylem tissue is continually produced toward the inside of the stem while more phloem tissue is continually produced toward the outside of the stem. This tissue is created in rings inside the developing and growing stems. These rings of tissue are usually produced in large quantities during the growing seasons of the year, and then very little growth occurs in these same regions for the rest of the year. These rings can be used to calculate the age of the plant, because the time interval when they are created is usually known. These rings can also be used to determine the climatic conditions the plant experienced during certain years of its life, because these growth rings are larger or smaller depending on the weather of the growing season.

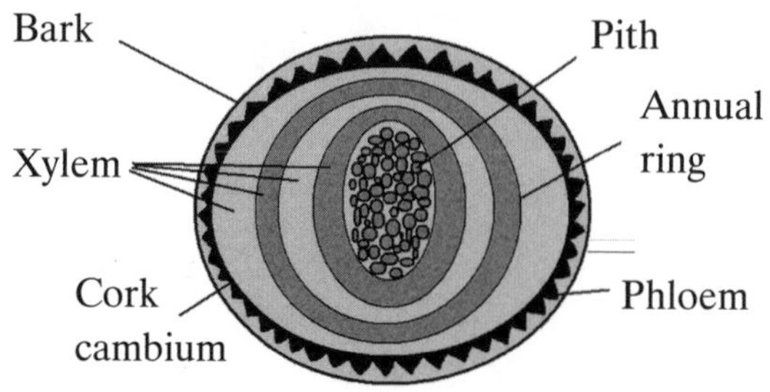

The cork cambium is found near the outer stem of the plant, close to the epidermis. This cambium layer also continually produces both xylem and phloem cells. The new cells produced from the cambium toward the inside of the stem become more xylem tissue, while the new cells produced toward the outside of the stem become more phloem tissue. These constantly produced phloem cells gradually push their way to the outside of the stem and eventually form the tough protective layer found on most trees and bushes called bark. The cells directly beneath the bark are alive and actively producing new protective layers, while the cells on the outer surface are collections of dead bark cambium, which will gradually slough off and be replaced by the dividing cells below them.

Growth of Roots

The tips of the roots of plants contain apical meristematic tissue. This tissue contains cells that are constantly dividing and, in turn, increasing the length of the root. Remember, as the root grows and pushes its way through soil, the root cap protects the ends of the root. Just as the arrangement of vascular bundles within the stems of plants differs between monocots and dicots, within the roots, the arrangement of vascular bundles also has distinct patterns between the two groups.

In the roots of dicot plants, the xylem is arranged in a star-shaped pattern at the center of the root. The phloem cells are found surrounding the xylem star.

In the roots of monocot plants, the xylem and phloem cells are found in alternating rings around the pith.

Practice Section 8-6

1. What is the name of the tip of the plant structure where there are actively dividing cells?

   a) pith b) cortex c) apical meristem d) lateral meristem

2. What section of the plant structure is responsible for producing the bark found surrounding most tree trunks?

   a) apical meristem c) xylem
   b) cork cambium d) vascular cambium

3. What protects the tip of the root as it grows and pushes its way through the soil?

   a) xylem b) phloem c) root cap d) root hairs

4. The meristematic sections of a plant constantly produce new cells from the process of

   a) mitosis b) meiosis c) conjugation d) germination

5. What is the name of the first seed leaf of a plant?

   a) epicotyl b) hypocotyls c) radicle d) cotyledon

6. Fertilization in a seed plant is achieved after the union of
   a) cotyledon and ovule
   b) pollen and ovule
   c) ovary and stigma
   d) style and pollen

Lesson 8-7: Plant Responses

When you hear a sudden loud noise, it may cause you to jump or scream. If you make too much noise while fishing in a pond, it will "scare" the fish away. If you swat at an annoying fly, it will try to dodge your attempt at hitting it. We are all very familiar with examples of animals responding to their environment. Something that is not as easy to see at first glance is how wonderfully plants also respond to their environment. Plants do not always respond in similar ways as animals to external stimuli, and this can lead to confusion about their ability to do so.

Hormones in Plants

A plant hormone is an organic compound synthesized in one part of the plant and moved to another, where it will cause a response to occur. Certain cells within the plant structure have receptors that will bind with particular hormones when they are released. The hormone that is released must be able to bind to these cells for the hormone to have action in that area of the plant. This is why when a particular hormone is released in one section of the plant only some areas in the rest of the plant are affected. These areas are the ones that contain the specific cells that will bind with that hormone. If the cells do not contain these "binding sites," they will not be affected directly by that hormone. Hormones regulate a range of activities within the plant including: growth, development, flowering, fruit ripening, and seed dormancy. There are five major classes of plant hormones.

Major Classes of Plant Hormones

Class	Effect on the Plant
Auxins	This hormone elongates cells in the shoot tips, but can also inhibit growth of lateral buds.
Gibberellins	This hormone stimulates the elongation and division of cells in the roots, seeds, and shoots. This hormone also causes the stimulation of flowering and seed germination.
Cytokinins	This hormone stimulates cell division in roots and fruits, and allows lateral buds to grow on stems.
Ethylene	This hormone, which is released as a gas, speeds up the ripening of fruit and also causes leaves and fruit to fall off the plant.
Abscisic Acid	This hormone prevents shoot growth and induces seed dormancy.

Table 8.9

Discovery of the First Plant Hormone

We have discussed the many contributions to the field of evolution by the scientist Charles Darwin earlier in this book. Charles Darwin is back again with his son Francis to help us a little more with the study of plants. In 1890, Charles Darwin and his son published their work on the experiments they had conducted on the growth of oat seedlings.

In their experiment, the Darwins grew oat seedlings under many different circumstances. When an oat seedling was left untouched, it would bend toward the light source in the room. This is a condition called phototropism that we will discuss later in this section. When another oat seedling had its tip removed, it did not bend towards light as it grew. When yet another oat seedling had its tip covered with an object that did not let light strike the tip of the plant, it also did not bend towards the light. This led Charles and Francis to assume that some unknown chemical inside the tip of these plants must be regulating the growth of the plant in response to light.

In 1926, Fritz W. Went was able to isolate the unknown chemical that the Darwins had predicted must be present. The chemical discovered by Went is called ***IAA*** (***indole acetic acid***). IAA is an example of a particular hormone called an auxin.

Botanists now believe that the chemical auxin is produced in the tips of plants. When light strikes one side of the plant stem, it causes auxin concentrations to increase on the shaded section of the stem. The increase in auxin in the shaded part of the stem causes the cells in the area to elongate. This elongation of the cells in the shaded part of the plant causes an unequal size distribution of cells in the stem, causing the stem to bend toward the light. It is important to note that that chemical auxin does not have the same effects in all parts of the plant. For example, in many plant roots, auxin actually causes the cells to grow slower.

Plant Movement

Although plants cannot move like animals in response to their environment, it has been shown that plants can indeed move parts of their structure through the process of hormonal growth. This type of directional growth of part of a plant in response to an external stimulus is called **tropism**. All tropisms in plants are named according to the stimulus that causes the growth. Table 8.10 shows some of the more common types of tropisms found within plants.

Tropism

Stimulus	Type of Tropism
light	phototropism
water	hydrotropism
chemical	chemotropism
gravity	geotropism
touch	thigmotropism

Table 8.10

If the growth movement of the plant is toward the stimulus, then the tropism is said to be *positive*. If the growth movement of the plant is away from the stimulus, then the tropism is said to be *negative*. For example, if the roots of a particular species of plant are negatively phototropic, then the roots will grow away from the light source; and if the stems of that same plant are positively phototropic, they will grow toward light.

Practice Section 8-7

1. Most plant responses are controlled by
   a) nerves b) hormones c) muscles d) all of these
2. If the roots of a plant are positively hydrotropic, then they will
   a) move toward water c) move toward light
   b) move away from water d) move away from light
3. Which of the following chemicals is used to help ripen fruit before it is sold in supermarkets?
   a) auxin b) abscisic acid c) ethylene d) adrenaline
4. Hormones control many activities within the plant structure. Which of the following activities is not controlled by hormones?
   a) growth c) fruit ripening
   b) flowering d) all of these are controlled by hormones
5. What was the first chemical hormone discovered in plants?
   a) abscisic acid b) ethylene c) IAA d) cytokinins
6. Auxin present in the roots of plants normally has which of the following effects?
   a) decreased cell growth c) no affect at all on the cells in the roots
   b) increased cell growth d) none of these

Lesson 8-8: The Fungi Kingdom

All fungi are multicellular (except yeast cells), nonmotile, heterotrophic organisms. Fungi obtain nutrition by absorption of nutrients through rootlike structures called ***rhizoids***. All members of the fungi kingdom are made of euakaryotic cells. The bodies of most fungi are composed of filaments called ***hyphae***. Many members of this group are parasitic, and many are important decomposers. Reproduction for members of this kingdom is sexual and asexual. The cells of fungi are surrounded by cell walls that are made of a substance called ***chitin***. There are more than 70,000 recognized species of fungi identified. Let us now look more closely at some of the fungal phylums in this group.

Phylum Ascomycota

The phylum Ascomycota is the largest group of fungi with more thanb 32,000 identified species. All members of this group contain a structure called an ***ascus***. The ascus is a sac-like structure that contains anywhere from four to eight haploid spores. Members of this group reproduce sexually and asexually. Members of this group are both saprophytic and parasitic. There are many species of this phylum that form symbiotic relationships with green alga to form ***lichens***. The alga makes the food for both partners and the fungi gathers water and minerals.

Yeast is the name given to one-celled members of this phylum. Many species of yeast fungi are used in baked products and alcoholic beverages. Some yeast fungal cells are pathogens of humans. Another example of Ascomycotes includes the very expensive, much sought-after truffles.

Phylum Basidiomycota

This phylum also is very large and contains more than 20,000 species. Members of this species are usually identified by the presence of a club-shaped, reproductive structure called the ***basidium***. In one of the better-known members of this phylum, the basidia are found in the gills under the cap of the fungal mushrooms, such as the species ***Amanita***.

Reproduction among members in the phylum is usually sexual, only the fungal forms of ***rusts*** and ***smuts*** reproduce asexually.

Phylum Zygomycota

There are about 1000 identified species in this phylum. The asexual spore-producing sporangia that sits on top of hyphae called sporangiophores identifies many members of this species. Sexual reproduction can also be accomplished by the touching of two different hyphae, which then grow together, swell, and produce haploid spores. Some species of this phylum live in the guts of arthropods in a symbiotic relationship. ***Rhizopus stolonifer***, usually found growing on old bread, is a well-known member of this phylum.

Parts of the Fungi

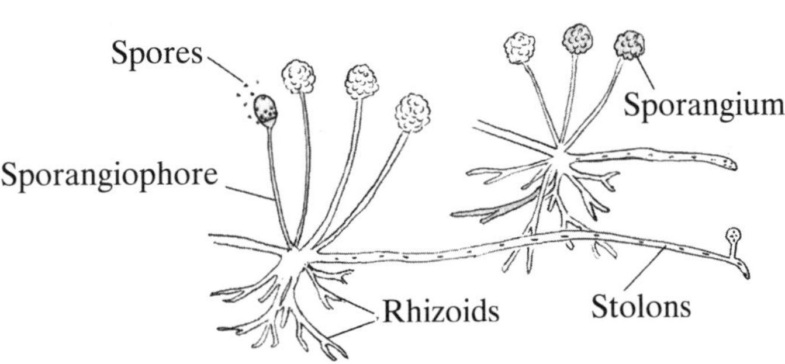

Practice Section 8-8

1. Members of the fungi kingdom
   a) are autotrophic
   b) are heterotrophic
   c) have structures composed of prokaryotic cells
   d) contain chlorophyll
2. Which of the following cell structures are found in fungi cells but not animal cells?
   a) cell membrane b) nucleus c) cell wall d) cytoplasm
3. The bodies of most fungi are composed of filaments called
   a) hyphae b) trichomes c) root hairs d) nematocysts
4. In what division would you find the mushroom-shaped species of fungi?
   a) Ascomycota b) Basidiomycota c) Zygomycota d) none of these
5. Members of the fungi kingdom reproduce
   a) asexually b) sexually c) both A and B d) neither A nor B
6. Some members of the fungi kingdom form symbiotic relationships with green alga to form
   a) yeast b) lichens c) mushrooms d) plants
7. What type of nutrition do members of the Fungi Kingdom perform?
8. What economic benefits do some fungi provide for humans?
9. What are lichens?

Lesson 8-9: The Protista Kingdom

This kingdom includes more than 60,000 species of usually one-celled (some can be multicellular), eukaryotic organisms. Asexual and sexual modes of reproduction can be observed among members of this kingdom. This kingdom is believed to have given rise to the animal, fungi, and plant kingdoms. The ancestors of this kingdom were the first eukaryotic cells that appeared on this planet about 1.5 million years ago. The members of this kingdom show great diversity in their modes of nutrition, life cycles, and structures. Scientists really have grouped many species into this kingdom not because of their similarities to other species within the group, but because they did not fit into any of the other established kingdoms. We will look more closely at some of the better-known phylums designated within this kingdom.

Phylum Sporozoa

Members of this phylum are unicellular, heterotrophic, nonmotile, and spore producing. The life cycles of species in this phylum are usually very complex and involve alternation of sexual and asexual reproduction. All members are parasitic; many are parasitic of animals, including humans. In fact, the protist ***Plasmodium***, which is responsible for the disease malaria, kills millions of humans every year. There are about 4,000 species within this phylum.

Phylum Rhizopoda

Organisms in the phylum Rhizopoda have irregular-shaped cellular bodies. Species in this phylum can move and capture food by projecting sections of their cell membrane outward from the cell body and filling these sections with cytoplasm. These cell-membrane projections are called ***pseudopods***. The word ***pseudopod*** is a Latin word that means "false foot." These cell membrane projections of members of this phylum acquired this name because when their movement is viewed under the microscope, it almost looks as if they are walking on feet. Many organisms normally observed in pond water under compound microscopes from this kingdom come from the genera ***Amoeba***.

Phylum Ciliophora

All organisms of this phylum are very highly structured members of Protozoa. All members of this phylum can reproduce sexually by conjugation, and asexually by binary fission. Organisms in the phylum Ciliophora achieve locomotion by structures called cilia. The most common member in this group is the ***Paramecium***, which lives in freshwater ponds.

Phylum Chlorophyta

This phylum consists of the protists called the green algae. The green algae can be unicellular, multicellular, or live in small colonies of cells. All members of this phylum are photosynthetic and are probably ancestors to members of the plant kingdom. Reproduction can be both asexual and sexual. There are about 7,000 species within this phylum that contain members such as ***Chlamydomonas***, ***Chlorella***, ***Volvox***, and ***Spyrogyra***.

Practice Section 8-9

1. The Protista kingdom is believed by many scientists to have given rise to which of the other following kingdoms?

   a) animal
   b) fungi
   c) plant
   d) all of these kingdoms

2. The disease malaria kills millions of people every year around the world. What is the name of the species from this kingdom responsible for this disease?
   a) *Chlamydomonas*
   b) *Chlorella*
   c) *Volvox*
   d) *Plasmodium*
3. What mode of nutrition is performed by organisms in the protista kingdom?
   a) autotrophic b) heterotrophic c) both A and B d) neither A nor B
4. All members of the protista kingdom of composed of
   a) eukaryotic cells
   b) prokaryotic cells
   c) A and B
   d) neither A nor B
5. Which phylum within the protista kingdom use cilia for locomotion?
   a) Sporozoa b) Rhizopoda c) Chlorophyta d) Ciliophora
6. *Pseudopod* is a Latin word that means
   a) circular body
   b) one-celled
   c) false foot
   d) containing chlorophyll

Lesson 8-10: The Bacteria Kingdoms

Bacterial cells usually are found in one of the following forms:

- Bacillus (rod-shaped cells)
- Coccus (spherical cells)
- Spirillum (spiral cells)

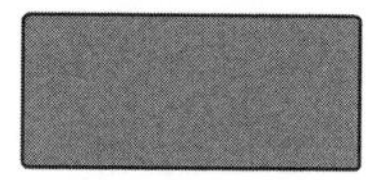
Bacillus

Coccus

Spirillum

• • • • • •

Bacteria are present as one-celled, chains or colonies of prokaryotes. The prokaryotes are the most abundant type of organism on the planet. The cells of bacteria lack membrane-bound organelles, which means that the DNA is loose within the cytoplasm of the cell. Reproduction can be accomplished asexually by the process of ***fission*** (a form of mitosis). Some species reproduce sexually by the exchange of genetic materials between two cells. Bacterial nutrition is by heterotrophic absorption of organic material, but in some species it is by photosynthesis and chemosynthesis. The many species of bacteria are very diverse in their habitats and their modes of obtaining energy. It is now recognized that prokaryotic cells actually form two different groups called the **Eubacteria** and the **Archaebacteria**. Even though both groups are prokaryotic organisms, they are so different chemically that they are now placed into two different kingdoms of bacteria. The major differences between the two bacterial kingdoms are summarized in Table 8.11.

Comparison of Bacterial Kingdoms

	Eubacteria	Archaebacteria
Cell Membrane	The cell membrane is made of normal phospholipids.	The cell membrane is composed of unique lipids.
Cell Wall	The cell walls are made of *peptidoglycan*.	Peptidoglycan is absent from the cell wall.
Gene Structure	The genes do not contain *introns*.*	The genes of some members of this kingdom contain *introns*.

* Introns are noncoding DNA sequences that are occur sometimes between coding sequences in the gene.

Table 8.11

Many members of the eubacteria kingdom are separated into various groups based on the chemical makeup of their cell walls. Bacteria that have large amounts of peptidoglycan in the cell walls, and when these walls are exposed to certain dyes becomes stained, are called ***gram-positive*** bacteria. Bacteria that has small amounts of peptidoglycan in the cell walls, and when these walls are exposed to certain dyes do not retain the stain, are called ***gram-negative*** bacteria. This knowledge of the cell walls of particular bacterial cells becomes a powerful tool in finding medicines for the treatment of certain bacterial infections.

There are probably about 10,000 identified species at this time, but it is very likely that this is only a small number of the total number of species that exist on this planet.

Cyanobacteria

Cyanobacteria are a group of bacteria that contain blue-green pigments. The mode of nutrition is the chemical process of photosynthesis. Reproduction is asexual in this group of bacteria. Members of this type of bacteria live in many types of environments, especially freshwater. Some examples of species found of this variety include ***Oscillatoria*** and ***Nostoc***.

Actinomycetes

Actinomycetes is a group of bacteria that has mycelial growth, and because of this is sometimes confused with fungi. This particular type of bacteria can be found in large numbers in most soil samples. They reproduce asexually by the production of spores. Members of this group produce most of the antibiotics used in medicine today such as streptomycin, actinomycin, and tetracycline. It is interesting to note that other members of this species can also be very harmful to humans, causing dental plaque, leprosy, and tuberculosis.

Spirochaetes

Members of this group are composed of long snakelike cells that can contain anywhere from two to more than 200 flagella inside the cell wall. Spirochetes are found in many habitats ranging from the ocean to the gastrointestinal tract of some animals. Species of spirochaetes that live in mud or water can withstand very low levels of oxygen. Members of this group are very mobile and can move very quickly through liquid mediums. Most spirochaetes are pathogens causing diseases such as syphilis and Lyme disease.

Practice Section 8-10

1. All bacteria are composed of
   a) eukaryotic cells
   b) prokaryotic cells
   c) A and B
   d) a noncellular structure
2. Bacteria complete the process of nutrition through the process of
   a) photosynthesis
   b) chemosynthesis
   c) heterotrophic absorption
   d) all of these
3. In recent years, bacteria found on this planet have been broken up into two separate kingdoms: eubacteria and archaebacteria. What is true about the differences between members of these two groups?
   a) Eubacteria are one-celled, while archaebacteria are multicelled.
   b) Archaebacteria perform heterotrophic absorption, while eubacteria perform photosynthesis.
   c) The cell walls of Eubacteria are made of *peptidoglycan*, while the cell walls of Archaebacteria do not contain this chemical.
   d) Eubacteria have no cell wall, while archaebacteria do have a cell wall.
4. Bacterial cells that have spherical shapes are given the name
   a) bacillus b) coccus c) spirillum d) none of these
5. A bacteria that has a cell wall composed of chemicals that readily absorb certain chemical lab dyes and become stained are called
   a) gram-positive bacteria
   b) gram-negative bacteria
   c) archaebacteria
   d) all of these
6. Which members of bacteria are used to produce most of the antibiotics used in medicine today?
   a) Cyanobacteria b)Actinomycetes c) Spirochaetes d) Archaebacteria

Terms From Chapter 8		
angiosperm	kingdom	root cap
archaebacteria	meristem	root hairs
bilateral symmetry	Nematocyst	sepal
binomial nomenclature	notochord	spongy cells
coelom	ovary	stamen
cortex	ovule	stigma
cotyledon	palisade cells	stomata
cuticle	pentaradial symmetry	style
dermal tissue	petal	taxon
epidermal cells	phloem	taxonomy
eubacteria	pistil	transpiration
gram-negative bacteria	pith	trichomes
gram-positive bacteria	phylum	tropism
ground tissue	pollination	vascular tissue
guard cells	radial symmetry	vertebral column
hyphae	rhizoid	xylem

Chapter 8: Exam

Matching Column for Kingdoms of Life

Match the description of the kingdom or an organism from that kingdom to the proper classification within the six-kingdom system. You may use an answer choice more than once or not at all.

Description or organism

1. Organisms in this kingdom are composed of prokaryotic cells.
2. Mushroom.
3. This kingdom contains multicellular, photosynthesizing organisms.
4. Lichen.
5. *Homo sapien*.
6. All the organisms in this kingdom are heterotrophic and are composed of cells that do not contain cell walls.
7. One-celled organisms that live in extreme environments such as near deep-ocean vents or in hot pools.

Kingdom

a. animal
b. plant
c. protista
d. fungi
e. eubacteria
f. archaebacteria
g. A and B
h. C and D
i. E and F

Matching Column for Structures in Flowering Plants

Match the structure found in a flowering plant with its proper function. You may use an answer choice more than once or not at all.

Structure in the plant

8. petal
9. stigma
10. ovule
11. pollen
12. ovary
13. cotyledon

Function

a. The structure found inside the ovary that contains the plant's female chromosomes and will develop into a seed if fertilized.

b. Found inside the seed, and sometimes called the "seed leaf." This structure provides food for the developing plant embryo.

c. The flower leaves that attract insects and birds through the use of color, odor, and sugar for the process of fertilization.

d. The structure in the flower that contains the ovules.

e. The sticky structure found at the tip of the pistil that collects pollen for fertilization.

f. The tiny structure produced by the anther of the flower that contains the male chromosomes of the plant.

Multiple Choice

14. The scientific name of the white oak is *Quercus alba*. Which of the following organisms is most closely related to the white oak?
    a) *Felis domesticus*
    b) *Lumbricus terrestris*
    c) *Quercus rubra*
    d) *Passer domerticus*

15. If two organisms are in the same *genus*, they could be in different
    a) species
    b) families
    c) kingdoms
    d) orders

16. An organism that is multicellular, absorptive heterotrophic, and reproduces by spores is most likely found in the
    a) bacteria kingdom
    b) plant kingdom
    c) protista kingdom
    d) fungi kingdom

17. What phylum in the animal kingdom contains organisms with notochords?
    a) Echinodermata
    b) Arthropoda
    c) Chordata
    d) Cnidaria

18. Vascular tissue is used by many plants to conduct water and food throughout the structure of the plant. What plant *division* contains plants without a vascular system?
    a) Bryophyta b) Anthophyta c) Coniferophyta d) Pterophyta

19. What is the organelle found in plant cells that contains the green pigment chlorophyll?
    a) cell wall b) chloroplast c) vacuole d) centriole

20. What is the name of the storage cells found inside the vascular bundles of dicot stems?
    a) cortex b) epidermal c) pith d) guard cells

Short Response

21. When a green plant is placed in a sunny window for a period of time it will begin to grow towards the sunlight entering from the window. This process is called phototropism. Explain how the process of phototropism works in this plant.

22. In the plant division Anthophyta there are two classes of plants called *Monocotyledonae* and *Dicotyledonae*. Explain the major differences between these two classes of plants. Give one example of a plant from each class.

23. All living organisms on this planet have what is called a common name and then a scientific name that is created by the binomial nomenclature system. Give two advantages of using the scientific name over the common name for the organism.

24. Explain how the guard cells located on the leaves of plants conserve water.

25. Plants grow in a different manner than most other organisms on this planet. One key to these unique growth patterns is meristematic tissue. What are the types of meristemic tissue and how do they allow for different types of growth in plants?

Answer Key

Answers Explained Section 8-1

1. **C** is correct because at the present time, about 2 million organisms have been scientifically classified on the planet. Many scientists believe about 10–100 million more may exist that have not yet been classified.

2. **C** is correct because binomial nomenclature is a scientific system of naming organisms based on their genus and species.

   A, **B**, and **D** are not correct because kingdom, phylum, class, order, and family are other taxons in the classification system but not part of the binomial name.

3. **D** is the correct binomial nomenclature name for the human being with the proper notation is *Homo sapien*. The genus name is capitalized, the species name is lowercase, and the words are italicized.

   A is not correct because in *Homo Sapien*, the species name is capitalized. Only the genus name should be capitalized.

   B is not correct because in *homo Sapien*, the genus name is not capitalized and the species is capitalized. Only the genus name should be capitalized.

   C is not correct because in Homo sapien, the name is not italicized.

4. **C** is the correct order of taxons from most general to most specific (kingdom, phylum, class, order, family, genus, and species).

   A, **B**, and **D** are not correct because theses choice are not in decreasing order from most general to most specific.

5. **C** is correct because the kingdom that contains organisms made of prokaryotic cells is the monera ,or bacteria, kingdom.

   A, **B**, and **D** are not correct because members of the fungi, animal, plant, and protista kingdoms are made of eukaryotic cells.

6. **E** is correct because the most closely related group of organisms is the group that consists of *Canis familiaris* and *Canis lupus.* This can be deduced because both organisms are found in the same genus called *Canis*.

   A, **B**, **C**, and **D** are not correct because all the pairs in these groups contain organisms that have different species and genus names.

7. As compared to Eubateria, Archaebacteria have a cell membrane composed of unique lipids, peptidoglycan is absent from their cell walls, and the genes of some members of the Archaebacteria contain structures called *introns*.

8. Some examples of human diseases caused by bacteria include; leprosy, tuberculosis, syphilis and Lyme disease.

9. Bacteria are usually found in one of the following three body shapes; rod shaped (bacillus), spherical shaped (coccus), and spiral shaped (spirillum)

Answers Explained Section 8-2

1. **D** is correct because it is not true that all members of the animal kingdom are unicellular, in fact most are multicellular organisms.

   A, **B**, and **C** are not correct because all members of the animal kingdom are composed of eukaryotic cells, produce a protein called collagen, and are heterotrophic.

2. **E** is correct because humans, earthworms, jellyfish, and slugs are all members of the animal kingdom.

3. **C** is correct because the Annelida phylum has members with bilateral symmetry, a coelom, and well-developed circulatory, excretory, and nervous systems.

   A is not correct because members of the phylum Cnidaria have radially symmetrical bodies.

   B is not correct because members of the phylum Mollusca have distinguishing body features like a molluscan foot, mantle and radula feeding structure.

   D is not correct because members of the phylum Arthropoda have jointed legs and exoskeletons.

4. **C** is correct because humans are members of the Chordata phylum in the animal kingdom. Chordata means that all humans have a backbone.

   A is not correct because members of the phylum Mollusca have distinguishing body features like a molluscan foot, mantle, and radula feeding structure.

   B is not correct because members of the phylum Annelida have bilateral symmetry, a coelom, and well-developed circulatory, excretory, and nervous systems.

   D is not correct because members of the phylum Echinodermata have radial symmetry and are covered in "spiny skin."

5. **D** is correct because members of the phylum Chordata are distinguished by their vertebral column or backbone.

   A is not correct because an exoskeleton is one distinguishing feature of members of the phylum Arthropoda.

   B is not correct because pentaradial symmetry is one distinguishing feature of members of the phylum Echinodermata.

C is not correct because nematocysts is one distinguishing feature of members of the phylum Cnidaria.

6. **E** is correct because amphibians, reptiles, birds, and mammals are all members of the phylum chordata because they contain vertebral columns.

Answers Explained Section 8-3

1. **C** is correct because all members of the plant kingdom do not have reproductive structures called flowers.

   A, **B**, and **D** are not correct because all members of the plant kingdom are autotrophic, have cells that contain cellulose, and are multicellular.

2. **C** is correct because ferns are members of the division in the plant kingdom called Pterophyta.

   A is not correct because Anthophyta is the division of flowering plants.

   B is not correct because Bryophyta is the division of nonvascular mosses.

   D is not correct because Coniferophyta is the division of plants that reproduce with cones.

3. **D** is correct because the division in the plant kingdom that contains the most identified species is Anthophyta. The Anthophyta division contains about 240,000 species.

   A is not correct because Bryophyta division contains about 15,000 classified species.

   B is not correct because Pterophyta division contains about 12,000 classified species.

   C is not correct because Coniferophyta division contains about 500 classified species.

4. **C** is correct because alternation of generation in plants means that the plant alternates between a haploid and a diploid stage.

   A is not correct because all plants are autotrophic.

   B is not correct because some plants reproduce sexually and others asexually.

5. **D** is correct because chlorophyll is the green pigment used by plants to trap light energy to power the synthesis of organic molecules.

   A is not correct because chloroplasts are the organelles found in some plant cells that hold the pigment chlorophyll.

   B is not correct because chitin is a form of carbohydrate that makes up the cell walls of fungi and is also part of the exoskeleton of arthropods.

   C is not correct because cellulose is a carbohydrate that is found in the cell walls of plants.

6. **A** is correct because plants that are classified as angiosperms contain flowers as reproductive structures.

   B is not correct because pinecones are the reproductive case structures for plants in the division Coniferophyta.

   C is not correct because spores are the reproductive structures for plants in the division Pterophyta.

   D is not correct because the term "naked" seeds refers to the seeds found inside the pinecones of plants in the division Coniferophyta.

Answers Explained Section 8-4

1. **C** is correct because the chloroplast is only found in plant cells. Animal cells would not contain chloroplasts because animal cells do not produce their own organic molecules.

   A and **D** are not correct because vacuoles and cell membranes are found in both plant and animal cells.

   B is not correct because centrioles are only found in animal cells.

2. **B** is correct because the parenchyma cells are mainly responsible for food production and storage.

   A is not correct because collenchyma cells serve mainly as a support structure for the plant.

   C is not correct because sclerenchyma cells serve as support structures and for make up of the vascular system in the plant.

   D is not correct because prokaryotic cells are cells that do not contain a nucleus. Plants are not made of this type of cell, only bacteria are.

3. **C** is correct because the vascular tissue is responsible for transport in the plant.

   A is not correct because dermal tissue is the protective layer of cells on the outside of the plant.

   B is not correct because ground tissue is the majority of the cells in a plant that are responsible for the tasks of food production, food storage, and support.

   D is not correct because epidermal cells are a form of dermal tissue.

4. **C** is correct because the "tube" cells found in the plant that transport food are called phloem.

   A is not correct because xylem is the collection of tubes in vascular plants that transports water and minerals.

   B is not correct because epidermal cells are found on the outside of the plant and serve as a protective layer.

D is not correct because trichomes are thin hairlike structures found extending from the epidermal cells of leaves and roots of some plants.

5. **C** is correct because the stomates are the openings on the bottom surface of leaves where most transpiration of water occurs in the plant.

   A is not correct because cuticle is the waxy layer that covers the epidermal cells and prevents the loss of water from the plant.

   B is not correct because guard cells are the cells located at the bottom surface of leaves that regulate the opening and closing of the stomates.

   D is not correct because trichomes are thin hairlike structures found extending from the epidermal cells of leaves and roots of some plants.

6. **B** is correct because root hairs that extend from the surface epidermal cells of the root greatly increase the surface area of the root structure for water and mineral absorption.

   A is not correct because stomates are the openings at the bottom surface of leaves that control the flow of gases into and out of the plant.

   C is not correct because guard cells are the cells located at the bottom surface of leaves that regulate the opening and closing of the stomates.

   D is not correct because mesophyll is the collection of cells found in the interior region of leaves that help the plant in food production.

Answers Explained Section 8-5

1. **C** is correct because the most common manner of reproduction found in the plant kingdom is the use of flowers. More plants use flowers to reproduce than any other mechanism.

   A is not correct because spores are used by some plants, such as ferns, but this is still a small percentage of the total number of plant species.

   B is not correct because pinecones are used by some plants, such as junipers, but this is still a small percentage of the total number of plant species.

   D is not correct because binary fission is a reproductive method of bacterial cells not plants.

2. **D** is correct because the only female structure listed is the ovule. The ovule is found inside the ovary of the plant and becomes the seed after fertilization.

   A is not correct because sepals are neither male nor female structures. Sepals are the leaves that enclose the flower petals before the flower opens.

B is not correct because the anther is a male structure in the flower that produces the plants male sex cells called pollen.

C is not correct because the filament is a male structure that holds the anther up from the base of the flower.

3. **C** is correct because pollen is produced in the anther of the flower.

   A is not correct because pollen does not contain the diploid number of chromosomes. Pollen contains the haploid or monoploid number of chromosomes.

   B is not correct because pollen is produced in the anther, which is the male part of the flower.

   D is not correct because pollen is the male sex cell of the plant. The ovules are produced by the female part of the plant called the ovary.

4. **C** is correct because the part of the seed plant that becomes the fruit is the ovary.

   A is not correct because pollen is the male sex cell that fertilizes the ovules of the flower.

   B is not correct because the ovules become the seeds of the plant that will be surrounded by the ovary that has developed into fruit.

   D is not correct because the sepals are the leaves that surround the petals of the flower before the flower opens.

5. **C** is correct because in the class *Monocotyledonnae*, the plants all have parallel veins in their leaf structures.

   A is not correct because the class *Dicotyledonnae*, not the class *Monocotyledonnae*, contains two seeds leaves.

   B is not correct because flower parts are in multiples of fours and fives in the class *Dicotyledonna*; in the class *Monocotyledonnae*, they are in multiples of three.

   D is not correct because vascular tissue is arranged in rings inside the stems of plants in the class *Dicotyledonna*; in the class *Monocotyledonnae*, the vascular tissue is scattered throughout the stem's interior.

6. **C** is correct because an oak tree is a member of the *Dicotyledonna* class. The oak tree contains leaves that have veins branched throughout the interior.

   A, **B**, and **D** are not correct because palm trees, lawn grass, and orchids are all members of the *Monocotyledonnae* class. The easiest to identify trait of their classification is that all of these plants have parallel veins in their leaves.

Answers Explained Section 8-6

1. **C** is correct because the tip of a plant where there are actively dividing cells is called the apical meristem.

   A is not correct because the pith is the region of parenchyma cells in the interior of the stems and some roots that serve as a food storage center.

   B is not correct because the cortex is the region of parenchyma cells found directly below the epidermal cells of the plant.

   D is not correct because the lateral meristem is the region of actively dividing cells found near the vascular tissue in the plant and the cork cambium. Division of cells in these regions allows the plant to grow in width.

2. **B** is correct because the cork cambium is the section of the plant that is responsible for producing the bark found surrounding the tree's trunk.

   A is not correct because the apical meristem is the tip of a plant where there are actively dividing cells.

   C is not correct because the xylem is the collection of tissue responsible for the transport of water and minerals throughout the plant.

   D is not correct because vascular cambium is the region of tissue responsible for producing more vascular tissue inside the plant.

3. **C** is correct because the root cap protects the tip of the root as it grows and pushes its way through the soil.

   A is not correct because xylem is the collection of tissue responsible for the transport of water and minerals throughout the plant.

   B is not correct because phloem is the collection of tissue responsible for the transport of food throughout the plant.

   D is not correct because root hairs are extensions of the epidermal cells of the root structure that increase the surface area of the root.

4. **A** is correct because the constantly reproducing cells of the meristematic tissues of a plant divide by the process of mitosis.

   B is not correct because meiosis is the process by which sex cells within an organism divide and reproduce.

   C is not correct because conjugation is a type of sexual reproduction where two unicellular organisms exchange genetic information.

   D is not correct because germination is the early growth stage of the plant embryo.

5. **D** is correct because the first seed leaf of a plant is called the cotyledon.

A is not correct because the epicotyl is the part of the plant growth above the cotyledon.

B is not correct because the hypocotyls is the part of the plant growth below the cotyledon.

C is not correct because the radicle is the first part of the plant to emerge from the seed during germination.

6. **B** is correct because fertilization is the union of pollen and ovule in a plant.

   A is not correct because cotyledon is the first seed leaf of a plant and the ovule is the female sex cell of the plant.

   C is not correct because the ovary is the female structure of the plant that produces the ovules and the stigma is the tip of the female structure of the plant that receives the pollen.

   D is not correct because style is the female structure of the plant that holds the stigma above the ovary and the pollen is the male sex cell of the plant.

Answers Explained Section 8-7

1. **B** is correct because most plant responses are controlled by hormones.

   A is not correct because nerves are not found in plants. Only animals contain nerves.

   C is not correct because muscles are not found in plants. Only animals contain muscles.

2. **A** is correct because if a plant's roots are positively hydrotropic, the roots will grow (move) toward water.

   B is not correct because if a plant's roots move away from water they are negatively hydrotropic.

   C is not correct because if a plant's roots move toward light, they are positively phototropic.

   D is not correct because if a plant's roots move away from light, they are negatively phototropic.

3. **C** is correct because the chemical ethylene is used to prematurely ripen fruit that is sold in supermarkets.

   A is not correct because auxin is a plant hormone that causes cells in the shoot tips to elongate and lateral buds not to grow.

   B is not correct because absciscic acid is a plant hormone that induces seed dormancy and prevents shoot growth.

D is not correct because adrenaline is a hormone found in animals, not plants. Adrenaline causes the heart and lungs to work at a faster rate.

4. **D** is correct because all of the answers are controlled by hormones in the plant: growth, flowering, and fruit ripening.

5. **C** is correct because the chemical IAA was the first plant hormone discovered; and the discovery took place in 1926 by Fritz Went.

   A, **B**, and **D** are not correct because abscisic acid, ethylene, and cytokinins were discovered in the years that followed this groundbreaking work by Went.

6. **A** is correct because many studies have shown that auxin actually decreases cell growth in the roots of plants.

   B is not correct because auxin causes increased cell growth in the shoots of plants.

   C is not correct because auxin usually has no effect on cells that do not have the target proteins located on their surfaces.

Answers Explained Section 8-8

1. **B** is correct because members of the fungi kingdom are heterotrophic. Fungi absorb organic materials into their cells through structures called rhizoids.

   A is not correct because fungi are not capable of making their own food.

   C is not correct because bacteria are made of prokaryotic cells; fungi are composed of eukaryotic cells.

   D is correct autotrophic organisms, such as plants, contain chlorophyll to help them trap light energy to make food. Fungi do not contain this chemical.

2. **C** is correct because all fungi cells are surrounded by a cell wall made of chitin. Animal cells do not have a cell wall.

   A, **B**, and **D** are not correct because the cell membrane, nucleus, and cytoplasm are all found in both animal and fungi cells.

3. **A** is correct because the bodies of most fungi are composed of filaments called hyphae.

   B is not correct because trichomes are the hairlike extensions of the epidermal cells of a plant and root.

   C is not correct because root hair is a type or trichome that helps increase the surface area of the root structure.

   D is not correct because nematocysts are the stinging cells found at the tip of the tentacles of members of the phylum Cnidaria.

4. **B** is correct because the species of mushroom-shaped fungi are found in the division Basidiomycota.

   A is not correct because Ascomycota is a division of fungi whose species all contain the reproductive structure called the *ascus*. The ascus contains the reproductive spores of the organism.

   C is not correct because Zygomycota is a division of fungi whose species are identified by the asexual spore producing sporangi.

5. **C** is correct because the fungi kingdom contains species that produce asexually and some species that produce sexually.

6. **B** is correct because some members of the fungi kingdom in the division Ascomycota form symbiotic relationships with green alga to form an organism called lichen.

   A is not correct because yeast is a type of one-celled fungi present in the division Ascomycota.

   C is not correct because mushrooms are type of fungi containing the reproductive structures called basidiums present in the division Basidiomycota.

   D is not correct because plants are organisms that are autotrophic and multicellular that are found in their own kingdom.

7. All members of the Fungi Kingdom perform heterotrophic nutrition. Fungi usually eat organic materials and act as decomposers.

8. Fungi are used in the production of alcoholic beverages, many baked products like bread, and cheese. Some species of fungi can be eaten directly like non-poisonous varieties of mushrooms.

9. Lichens are living organisms formed by an interaction between fungi and algae. The algae makes the food for both organisms and the fungi absorbs the water and minerals needed by both. This is considered to be a mutualistic partnership.

Answers Explained Section 8-9

1. **D** is the correct answer because most scientists now believe that the protista kingdom gave rise to the animal, fungi, and plant kingdoms found on this planet.

2. **D** is the correct answer because the protist *Plasmodium* is the organism that causes the disease malaria. This protist is transported by the *Anopheles* mosquito from host to host.

   A is not correct because *Chlamydomonas* is a type of green alga found in the protista kingdom.

B is not correct because *Chlorella* is a type of green alga found in the protista kingdom.

C is not correct because *Volvox* is a type of green alga found in the protista kingdom.

3. **C** is the correct answer because the protista kingdom contains some organisms that are autotrophic, some organisms that are heterotrophic, and some organisms that are both autotrophic and heterotrophic.
4. **A** is the correct answer because all members of the protista kingdom are composed of eukaryotic cells.

   B is not correct because all members of the bacterial kingdom are composed of prokaryotic cells.
5. **D** is the correct answer because the Ciliophora phylum in the protista kingdom has members that use the structures called cilia for movement.

   A is not correct because members of the Sporozoa phylum do not have structures for locomotion.

   B is not correct because members of the Rhizopoda phylum move by means of pseudopods.

   C is not correct because most members of the Chlorophyta phylum are nonmotile.
6. **C** is the correct answer because the word *pseudopod* in Latin means "false foot." Pseudopods are structures that members of the phylum Rhizopoda use for locomotion.

Answers Explained Section 8-10

1. **B** is the correct answer because all members of the bacteria kingdom are composed of prokaryotic cells.

   A is not correct because all members of the other kingdoms of life on Earth are composed of eukaryotic cells.
2. **D** is the correct answer because the bacteria kingdom contains some species that complete the process of nutrition by photosynthesis, some species by chemosynthesis, and some species by heterotrophic absorption of food.
3. **C** is the correct answer because the difference between the two bacterial kingdoms is that members of the eubacteria have cell walls made of peptidoglycan, while the cell walls of archaebacteria do not contain this chemical.

   A is not correct because all bacteria in both kingdoms are unicellular.

B is not correct because in both kingdoms of bacteria there are many modes of nutrition.

D is not correct because both kingdoms of bacteria contain cell walls.

4. **B** is correct because the groups of bacterial cells circular in shape are called *coccus* bacteria.

   A is not correct because bacillus bacteria are rod-shaped.

   C is not correct because spirillum bacteria are spiral in shape.

5. **A** is correct because gram-positive bacteria have cell walls composed of chemicals that readily absorb certain chemical lab dyes and become stained.

   B is not correct because gram-negative bacteria have cell walls that do not readily absorb certain chemical dyes.

   C is not correct because this chemical stain test is not done on members of the archaebacteria kingdom because of their chemically different cell wall structures to that of members of the eubacteria kingdom.

6. B is correct because most of the antibiotics used in medicine today are produced from members of the Actinomycetes phylum of bacteria.

   A is not correct because Cyanobacteria are autotrophic, blue-green bacteria.

   C is not correct because Spirochaetes are very motile, disease-causing members of the bacteria kingdom.

   D is not correct because archaebacteria are a distinct kingdom of bacterial cells that have genetic properties more similar to eukaryotic cells than eubacterial cells.

Answers Explained Chapter 8 Exam

Matching Column

1. I	5. A	10. A
2. D	6. A	11. F
3. B	7. F	12. D
4. H	8. C	13. B
	9. E	

Multiple Choice

14. **C** is the correct answer because *Quercus rubra* is in the same genus *Quercus* as the white oak *Quercus alba.*

    A, **B**, and **D** are not correct because it can be seen that the organisms in these choices: *Felis domesticus*, *Lumbricus terrestris*, and *Passer domerticus* are all found in different genera than *Quercus alba*.

15. **A** is correct because two organisms can be in different species but in the same genus of classification.

    B, **C**, and **D** are not correct because if two organisms are classified into the same genus they must also be in the same family, kingdom, and order.

16. **D** is correct because most members of the fungi kingdom reproduce with spores, are multicellular and are absorptive heterotrophs.

    A is not correct because members of the bacteria kingdom are unicellular and don't reproduce using spores.

    B is not correct because members of the plant kingdom are autotrophic in their mode of nutrition.

    C is not correct because members of the protista kingdom are usually unicellular and do not reproduce by spores.

17. **C** is correct because members of the phylum Chordata all contain a structure called a notochord.

    A, **B**, and **D** are not correct because members of the phylums Echinodermata, Arthropoda, and Cnidaria do not contain the structure called the notochord.

18. **A** is correct because members of the plant division Bryophyta do not have developed vascular systems to transport materials throughout their structures.

    B, **C**, and **D** are not correct because the members of the plant divisions Anthophyta, Coniferophyta, and Pterophyta all contain vascular tissue.

19. **B** is correct because the organelle found in plant cells that contains the green pigment chlorophyll is the chloroplast.

    A is not correct because the cell wall is the clear, rigid organelle that gives the plant cell shape and support.

    C is not correct because a vacuole is the organelle found in plant and animal cells used for storage of food and sometimes waste.

    D is not correct because a centriole is an organelle only found in animal cells and used in cell division.

20. **C** is correct because the name of the storage cells found inside the center of the stems of dicot plants is called pith.

    A is not correct because cortex is the storage cell area located outside of the vascular tissue, below the epidermal cells of many plant stems.

    B is not correct because epidermal cells are located on the surface of the stems, roots, and leaves, and provide a protective barrier for the inner cells of the plant.

D is not correct because guard cells are located around the stomata located on the epidermal cells on the bottom of the leaves.

Short Response

21. At the tip of the plant stem is an area called the apical meristem. The cells in this region produce a hormone called auxin. Cells that are exposed to the chemical auxin grow and elongate to a greater extent than cells that are not exposed to this chemical. When a plant is placed in the sunlight, only one side of the plant stem is exposed to the sun. Sunlight destroys the auxin in these cells, but the cells on the shady side of the stem continue to grow at an accelerated rate. This causes the shady side of the stem to outgrow the sunny side of the stem, which, in turn, causes the stem to bend toward the light.

22. The following chart shows all the major differences between the two classes of plants found in the Anthophyta division of plants.

Characteristics of Monocots and Dicots

	Monocots	**Dicots**
Leaves	Parallel veins within the leaves	Branched veins within the leaves
Flowers	Flower parts (petals, sepals, stamens, or pistils) usually in multiples of three	Flower parts (petals, sepals, stamens, or pistils) usually in multiples of four or five
Stems	Vascular tissue scattered throughout the inside of the stem	Vascular tissue arranged in a ring inside the stem.
Seed leaves	Contains a single seed leaf or cotyledon	Contains two seed leaves or cotyledons
Roots	Root structures are fibrous	Root structure is a taproot

Table 8.12

An example of a monocot would be an orchid plant and an example of a dicot would be an oak tree.

23. Two advantages of using the scientific name of the organism instead of the common name of the organism are:

    a. The scientific name is always written in Latin. Latin is a language used by all scientists who classify organisms around the world. There is no language barrier to overcome when sharing data about the organism's name.

    b. One particular organism can have many different common names, even in the same country. An organism will always only have one scientific name. If you mention the common name, someone may not know the exact species you are speaking about, but if you use the scientific name, there will be no confusion.

    c. Guard cells are located around the openings on the leaves called stomata. When it is very warm outside the plant loses large amounts of water by transpiration through these openings called stomata. To help conserve water, the guard cells close the stomata on these very hot days by swelling and covering these openings.

    d. There are two major types of meristematic tissue: apical meristematic tissue and lateral meristematic tissue. Apical meristematic tissue is located at the tips of the roots and stems of plants. The apical meristematic tissue has actively dividing cells that enable the plant to grow longer and taller.

       Lateral meristematic tissue comes in two forms: vascular cambium and cork cambium. These two types of tissue contain actively dividing cells that help thicken the plant's stem or increase its width.

GLOSSARY

Abiotic factor: A nonliving influence found within the ecosystem.

Acid: A chemical that contains hydrogen atoms and when dissolved in a liquid releases hydrogen ions.

Activation energy: The quantity of energy required to start a chemical reaction.

Active transport: The movement of molecules from an area of low concentration to an area of high concentration with the expenditure of cell energy.

Adenosine diphosphate (ADP): The molecule that results from the removal of a phosphate group from an ATP molecule.

Adenosine triphosphate (ATP): The most available energy source for living cells produced during cellular respiration.

Adhesion: Attractive forces between molecules of different substances.

Adrenaline: A hormone that is secreted by the adrenal gland that activates the body during stressful situations.

Aerobic respiration: The process of respiration with the use of oxygen.

Allele: The alternate genes for a particular trait found at the same spot on two different chromosomes.

Alternation of generations: When an organism alternates between sexual and asexual reproductive stages.

Alveoli: The area of the lungs where gas exchange between the environment and the body tissue occurs.

Anaerobic respiration: The chemical process of oxidizing food molecules for the release of energy in the absence of oxygen.

Analogous structures: Structures found in two organisms that have the same function but are structurally different and come from different origins.

Angiosperms: Plants that reproduce using the flower that produces seeds and fruit.

Artery: The thick muscular blood vessel that brings blood under high pressure away from the heart and to the body tissue.

Asexual reproduction: Reproduction of a new cell or organism from one parent or one set of genetic information.

Atom: The smallest particle of an element that retains the properties of that element.

Atomic number: The number of protons in an atom that determines the nature of the element.

Autotroph: An organism that has the ability to make its own organic food molecules from inorganic molecules.

Auxin: A plant hormone that promotes of inhibits cell growth.

GLOSSARY

B

Base: A chemical that when dissolved in a liquid increases the hydroxide concentration of the solution.

Binomial nomenclature: The scientific name that consists of the genus and the species of the organism.

Biome: A large geographic section of the planet that is named according to the climax community that is the most prevalent.

Biosphere: The entire surface of the planet that contains all the land, air and water necessary for life.

Biotic factor: A living organism that influences other living organisms within an ecosystem.

Bronchi: The two tubes that extend from the trachea to each lung.

Bronchioles: The smaller tubes that extend from the bronchi into the lungs.

C

Cambium: A meristematic plant tissue that contains constantly dividing cells that increase the diameter of the stems and roots.

Capillary: The blood vessel in the body with a very small diameter where most diffusion of materials between the circulatory system and the body tissue occurs.

Carbohydrate: A macro-organic molecule containing the element carbon and the elements hydrogen and oxygen in a 2:1 ratio. Many organisms use carbohydrates as a source of quick energy.

Carnivore: A meat-eating organism.

Catalyst: A substance that affects the speed of a chemical reaction without itself being destroyed in the process.

Cell: The basic unit of structure and function of all living organisms.

Cell membrane: A phospholipid bilayer with embedded proteins that controls the flow of materials into and out of the cell.

Cellulose: A carbohydrate that is found in the cell walls of plant cells.

Cell wall: A rigid wall that surrounds plant and bacterial cells. This organelle provides structure and support while also allowing the movement of materials across its surface.

Central nervous system: The brain and the spinal cord.

Centriole: An organelle involved in the formation of spindle fibers used in cell division in animal cells.

Cerebellum: The section of the brain that controls balance, coordination, and voluntary muscle movement.

Cerebrum: The section of the brain that controls thought, memory, creativity, and reasoning in human beings.

Chemical bond: A strong force of attraction holding atoms together in a molecule.

Chlorophyll: A green pigment used by some cells to trap energy from light to power the process of photosynthesis.

Chloroplast: An organelle found in photosynthetic cells that contains the green pigment chlorophyll.

Chromatography: A method of separating chemicals based on their deposition rates on a piece of paper.

Chromosome: A structure found within the nucleus of a cell containing the genetic material.

Climax community: A stable ecological community that usually is the last stage during ecological succession.

Codominance: The expression in the phenotype of two alleles of a heterozygous pair.

Codon: Three nitrogen bases on the messenger RNA molecule that determine a particular amino acid.

Cohesion: The attractive forces between molecules of the same substance.

Commensalism: A symbiotic relationship where one organism benefits and the other organism is not affected.

Community: All the plants and animals that interact in a particular location.

Competition: A condition caused when one organism requires the same food and or shelter as another organism at the same time.

Compound: Two or more elements that are chemically combined.

Compound microscope: A microscope that uses two lenses to magnify and focus on small objects. This microscope allows for magnifications in the range of 2-10,000 times.

Condensation: The phase change of water vapor changing into liquid water.

Consumer: An organism that feeds on producers.

Control group: The part of the experiment that is kept constant to use as a basis of comparison for the variable.

Cotyledon: The seed leaf that feeds the developing plant embryo.

Covalent bond: A bond formed between atoms by the sharing of electrons.

Crossing over: The process where genetic information is exchanged between homologous chromosomes during the meiosis. Linked groups of genes are usually separated during this process.

Cross-pollination: Transfer of pollen from one plant to another kind of plant.

Cytoplasm: A jelly-like clear fluid that fills the entire volume of the cell and provides the environment needed for many of the chemical reactions that occur in the cell.

Data: *See quantitative or qualitative data.*

Dehydration synthesis: A chemical reaction where smaller molecules are joined to create larger more complex molecules with the loss of water in the process.

Denitrification: The process of nitrates in the soil being broken down by bacteria with nitrogen gas being released as a product into the atmosphere.

Diaphragm: A sheet of muscle that separates the chest cavity from the abdominal cavity and functions in breathing.

Diffusion: The random movement of molecules from an area of high concentration to an area of low concentration.

Digestion: The process of breaking large complex organic molecules into smaller less complex organic molecules by physical and chemical actions.

Diploid: The full number of chromosomes for the species in the cell.

Disease: A condition in a living organism where one or more biological systems are not working properly.

DNA: (deoxyribose nucleic acid) Genetic material that is responsible for the replication of life.

DNA polymerase: An enzyme that will split a DNA molecule along the hydrogen bonds that connect the nitrogen bases within the molecule. This action must occur during DNA replication.

Dominant gene (trait): The gene that expresses itself no matter what other gene is matched with it during fertilization.

E

Ecology: The study of relationships between living things and their environment.

Ecological succession: The gradual change of the plant environment of a particular ecosystem over time.

Ecosystem: All the living and non-living factors found in a particular location and their interactions.

Egestion: The removal of undigested food from the body.

Egg: The female gamete or ovum.

Electron: A negatively charged subatomic particle that is found moving at high speeds around the nucleus of an atom.

Electron balance: A device used to electronically measure the mass of an object.

Electronegativity: The tendency for certain atoms to gain electrons and acquire a negative charge.

Electron microscope: A microscope that uses a beam of electrons instead of light to magnify objects. This microscope allows for magnifications in the range of 500,000 times.

Electrophoresis: A method of separating very small chemical set in a gel based on their charges.

Element: A substance that is made up of all of the same type of atom.

Elimination: The removal of undigested food from the body.

Embryo: A living organism in the early stage of its growth and development.

Endocrine gland: A gland that functions without ducts and secretes hormones into the bloodstream.

Endocytosis: A form of active transport where the cell membrane will actually surround the material forming a vacuole. The material never crosses the cell membrane in this process.

Endoplasmic reticulum: An organelle consisting of a series of membrane channels that run throughout the cell's interior providing space for specific chemical reactions.

Endoskeleton: A hard, internal skeleton.

Enzyme: An organic catalyst that speeds up the rate of chemical reactions within the cells and bodies of living organisms.

Epidermis: The thin cells that make up the outer layer of skin of a plant or animal.

Epiglottis: The flap of skin that covers and prevents food from entering the trachea during the action of swallowing.

Esophagus: The tube that connects the oral cavity to the stomach.

Estrogen: A female hormone responsible for secondary sex characteristics.

Eukaryote: A cell type that contains membrane bound organelles.

Evaporation: The phase change of liquid water becoming water vapor.

Evolution: The theory that living organisms that exist today have changed through time from earlier life that once lived on this planet.

Excretion: A life process that involves the removal of metabolic waste from the organism.

Exocytosis: A form of active transport where large particles are expelled from inside the cell.

Exoskeleton: A hard, external skeleton.

F

Fat: (lipid) A complex organic molecule used for energy storage in the body that is composed of three fatty acids and one glycerol.

Fermentation: The anerobic respiration reaction that produces small amounts of usable energy.

Fertilization: The fusion of the male gamete with the female gamete during the process of sexual reproduction.

Fibrinogen: A chemical in the body that aids in the clotting of blood.

Flower: The structure used for sexual reproduction in angiosperm plants.

Food chain: A focused look at a particular pathway through an ecosystem that shows what an organism eats (how it obtains energy) and what it is eaten by.

Food pyramid: A diagram that represents the amounts of energy found at each trophic level in an ecosystem.

Fossil: The remains or traces of an organism that lived in Earth's past.

G

Gamete: A sex cell, such as a sperm or egg.

Gene: The part of the DNA molecule that has the code for a particular piece of genetic information of the organism.

Gene frequency: The ratio of an allele to its other alleles in the gene pool.

Gene pool: The sum total of all the genes within a given population in study.

Genetics: The study of the processes involved when organisms pass on their genes to the next generation.

Genotype: The genetic makeup of the organism.

Germination: The early growth of new plant from a seed.

Glycolysis: The anaerobic process of respiration where glucose is oxidized into pyruvic acid and there is a small release of energy.

Golgi bodies: An organelle that prepares certain protein compounds for dispersal from the cell.

Gonads: The sex organs of an organism.

Greenhouse effect: An idea that large amounts of carbon dioxide being pumped into the planet's atmosphere could trap enough heat energy to raise the global temperature and disrupt climatic conditions around the world.

Growth: A life process that involves the increase in cell size and number within the organism.

H

Habitat: The physical location that an organism is best able to live.

Half-life: The amount of time is takes for half of the atoms of a sample of radioactive material to decay into a more stable form.

Hardy-Weinberg principle: This principal states that the frequency of particular alleles will not change in a given population under certain set conditions.

Hemoglobin: A chemical found in red blood cells that aids the cell in the transport of oxygen and carbon dioxide.

Herbivore: A plant-eating animal.

Heterotroph: An organism that is not able to synthesize organic molecules and so must look for them in the environment as a means of nutrition.

Heterozygous: When an organism has two different alleles for the same trait.

Homeostasis: The maintenance of a stable internal environment.

Hominid: A primate that walks on two feet and has a large brain case. The only surviving members of this group are humans.

Homologous structures: Structures found in two different organisms that have the same evolution and origins, but not the same function.

Homozygous: When an organism has two of the same alleles for a particular trait.

Hormone: Chemical that is secreted by a gland and travels to another target tissue within the body of the organism.

Hydrogen bond: A weak bond that forms between the atom hydrogen and electronegative atoms like oxygen.

Hydrolysis: A chemical reaction where large complex molecules are broken down into simpler smaller molecules by the addition of water.

Hypothesis: A possible answer to a problem that can be tested.

I

Incomplete dominance: A type of inheritance where two alleles are not fully expressed in the phenotype. Instead each allele is partially expressed causing a blending in the final phenotype.

Index fossil: The fossil of an organism that can be used to relatively date other fossils or rocks in the sediment.

Insulin: A chemical secreted by the pancreas that aids in the conversion of glucose to glycogen in the body.

Ion: An atom that has acquired a positive charge by losing electrons or an atom that has acquired a negative charge by gaining electrons.

Ionic bond: A bond formed between atoms by the transfer of electrons.

Isotopes: Atoms that have the same number of protons but different masses.

Karyotype: A diagram that shows the orderly arrangement of all the human chromosomes.

L

Lacteal: A extension of the lymph system in the villi of the small intestines that absorbs digested fat molecules.

Large intestine: The tube located in the gastrointestinal tract after the small intestine that is responsible for the absorption of water from the undigested food.

Ligament: A tissue that connects bone to other bone in the body.

Linkage: When genes that are present on the same chromosome stick together during cell division. Since the genes are inherited in groups the traits that are determined by these genes are also inherited in groups.

Lipid: A macro-organic molecule created by the union of three fatty-acid molecules with one glycerol molecule. Lipids are used by many organisms for long-term energy storage and insulation of the body.

Lymph: The liquid portion of the blood plasma that leaks out of the blood vessels into the surrounding body tissue that is collected by the lymphatic system.

Lysosomes: Small sac-like organelles that contain digestive enzymes that are used to dissolve food, and worn-out cell parts.

M

Matter: Anything that has mass and volume.

Medulla: The section of the brain that controls the basic processes of life such as breathing and the beating of the heart.

Meiosis: Cell division where the number of chromosomes is reduced in half.

Menstruation: The discharge of the uterine lining if the embryo does not implant.

Meristem: The tissue in a plant that consists of undifferentiated cells and where constant cell division occurs.

mRNA: (messenger RNA) An RNA molecule that brings instructions, for making proteins, from the DNA in the nucleus to the ribosomes in the cytoplasm.

Metabolism: All the chemical reactions of the life processes working together.

Metric system: A decimal system of measurement that uses powers of 10.

Microtubules: Tiny tube-like extensions that extend across the interior of the cell and act like a skeletal structure for the cell.

Mitochondria: An oval-shaped organelle that contains enzymes necessary for the chemical processes of respiration and energy production.

Mitosis: Cell division that results in two cells identical to the parent cell.

Mixture: The physical union of two or more substances.

Molecular formula: A formula that lists the type and number of atoms present in a molecule of a substance, usually in the lowest whole number ratio possible.

Molecular genetics: The study of the chemistry of the materials of inheritance such as DNA and RNA.

Monoploid (haploid): Half the number of chromosomes as the body cell of the species.

Monosaccharide: A simple sugar. Examples include glucose and fructose.

Mutation: A change in the DNA sequence that can, at times, be inherited.

Mutualism: A symbiotic relationship between two organisms where both organisms benefit from being in the relationship.

N

NAD: A coenzyme that transports high energy hydrogen atoms during the chemical process of respiration.

Natural selection: The idea that organisms that are best adapted to their environment will produce the most offspring for the following generation.

Nephron: The functional blood-filtering unit present in the kidney.

Neuron: A nerve cell.

Neurotransmitter: A chemical secreted by the end of nerve cells that continues the movement of the nerve impulse across the synapse between two nerve cells or between nerve cells and effectors such as muscles or glands.

Neutron: A neutral subatomic particle found in the nucleus of the atom.

Niche: The role a species plays in the ecosystem.

Nondisjunction: A chromosome mutation that occurs when two chromosomes fail to separate after the process of synapsis.

Nucleic acid: A macro-organic molecule consisting of alternating molecules of sugar and phosphates, with nitrogen bases attached to the sugar units. Examples of nucleic acids include DNA and RNA.

Nucleotide: The subunit of nucleic acids consisting of a nitrogen base, a phosphate group and a 5-carbon sugar.

Nucleus: A large circular organelle that controls all the chemical activities of the cell including reproduction.

Nutrition: A life process that involves the obtaining and use of food.

Organ: A group of tissue that performs a special function together in an organism.

Organelle: Structures found inside the cell that help the cell achieve the chemical processes necessary for life.

Organic compound: A usually very complex compound that contains carbon-hydrogen bonds.

Osmosis: The passive transport of water.

Ova: The reproductive egg of a female organism.

Ovary: The female reproductive organ.

Ovulation: The release of a mature egg from the follicle in the ovary.

Ovule: The structure in a flowering plant that is found inside the ovary and. if fertilized. will develop into the seed of the plant.

Oxidation: A chemical reaction where a compound loses hydrogen atoms or electrons and there is a release of energy.

Parasitism: A relationship between two organisms where one organisms benefits from the relationship while the other organism is harmed by the relationship.

Passive transport: The movement of molecules, without the expenditure of energy of the cell, from a region of high concentration to a region of low concentration.

Pathogen: Any organism that can cause disease.

Pedigree chart: A chart that shows some genetic connections between people in the same family.

Peptide: A chain of amino acids.

Peptide bond: The bond between two amino acids in a protein.

Periodic table: An orderly arrangement of all the known elements based on atomic number.

Peristalsis: The muscular contractions present in parts of the gastrointestinal tract that moves food along.

PGAL: (phosphoglyceraldehyde) The first organic compound produced during the chemical process of carbon fixation. PGAL is then used to synthesize other organic compounds.

pH: The measure of the acidity or alkalinity of a substance.

Phenotype: The appearance of an organism based on its genotype.

Phloem: A tissue that transports food throughout the structure of the vascular plants.

Phospholipids: Phosphorus and oxygen attached to lipid molecules. These chemicals make up the bi-layer of the cell membrane.

Photolysis: The chemical process by which a water molecule is broken into hydrogen and oxygen atoms by light energy.

Photosynthesis: The process where a cell uses energy from the sun to power a chemical reaction that creates organic molecules from inorganic molecules.

Platelets: Fragments of bone cells that aid the body in blood clotting.

Ploidy: A type of chromosomal mutation where too many chromosomes can end up in one of the developing cells.

Polysaccharide: A complex sugar consisting of polymers of simple sugars joined in long chains. Examples of a polymer include cellulose and starch.

Pollen: The male sex cell or gamete in sexually reproducing flowering plants.

Pollination: The transfer of the male reproductive structure called pollen to the female collecting structure called the stigma.

Polysaccharide: A complex carbohydrate.

Population: All members of a particular species that live in a specific location.

Predation: An action when one organism kills and feeds on another organism.

Producer: Organisms on the planet that are able to synthesize organic food molecules from inorganic molecules. In most cases these organisms use the sun as an energy source to complete this task.

Progesterone: A female hormone that aids in the build up of the uterine lining for pregnancy.

Protein: A macro-organic molecule created by long chains of amino acid molecules. Proteins are the organic molecules important in the structures of living tissue.

Proton: A positively charged subatomic particle found in the nucleus of an atom.

Punnett square: A method of determining the genetic probability between the cross of two organisms.

Qualitative data: Data collected that is descriptive in nature.

Quantitative data: Data collected that contains numbers.

Radial symmetry: An organism that can be divided into identical halves down the center of the organism.

Radioactive dating method: A method of dating fossils based on the decay of radioactive atoms.

Receptor sites: Specific tissue that a hormone will bind to and affect, within the body of an organism.

Recessive gene (trait): A gene that will only be expressed phenotypically when it is matched with another recessive gene.

Rectum: The storage site of feces located after the large intestines in the gastrointestinal tract.

Red blood cells: The cells in an organism that are responsible for the movement of oxygen and carbon dioxide throughout the body.

Regulation: A life process that controls and coordinates the various chemical and mechanical processes within the body of the organism.

Relative dating: A method of determining the age of fossils by comparing their placement in layers of sediment to other fossils buried above or below them.

Reproduction: The process by which a one celled to multicelled organism makes another of its own kind.

Reproductive isolation: Separation of populations so that they cannot breed and produce more offspring of their own kind.

Resolution: The ability of a lens to distinguish clearly between two objects that are very close together.

Resolving power: The ability of a microscope or some other viewing device to distinguish between two points that are close together.

Respiration: A life process that involves the oxidation of food molecules for the release of energy.

Ribosome: The smallest organelle in the cell that is the site of protein synthesis.

RNA (ribonucleic acid): A nucleic acid molecule that works with DNA and the genetic instructions of the organism.

Root hair: Hairlike extensions of the epidermal cells that help increase the surface area of the roots for water absorption.

S

Secondary consumer: A carnivore that feeds on plant-eating animals.

Scientific method: An organized approach to gathering information and solving a problem.

Sex chromosomes: The chromosomes that determine the sex of the organism.

Sex-linked trait: A trait that is controlled by a gene or genes on one of the sex chromosomes.

Sexual reproduction: A type of reproduction where there is a fusion of two sex cells or gametes.

Skeleton: The hard structures that give support and shape to the body of many organisms.

Small intestine: The tube located after the stomach in the gastrointestinal tract that is responsible for the bulk of chemical food digestion and absorption of food into the bloodstream.

Solute: The substance that is dissolved in the solvent making a solution.

Solution: A mixture with uniform makeup between a solute and a solvent.

Solvent: A liquid that dissolves another substance to make a solution.

Speciation: The process where a species becomes two or more different species.

Species: A group of organisms that can mate and produce fertile offspring.

Sperm cell: The male gamete or reproductive haploid cell.

Spontaneous generation: The belief that living organisms could arise from inorganic matter.

Spore: A small, reproductive cell that gives rise to new organisms.

Sporophyte: The spore producing stage in an organism.

Stamen: The male reproductive organ of the plant.

Stimulus: Any factor that triggers the dendrites of a nerve cell.

Stomach: The muscular organ in the human body responsible for the storage of food and the chemical digestion of proteins.

Stomata: The small openings found on the bottom surface of leaves where gases enter and leave the plant.

Structural formula: The structural formula of a substance shows how the atoms are connected and arranged within the molecule.

Substrate: The material acted upon by the enzyme.

Succession: The process by which one type of community is replaced by another in an environment.

Synapse: The microscopic space located between two nerve cells where neurotransmitters are released.

Synthesis: A life process that involves the chemical joining of two smaller molecules into one larger more complex molecule.

T

Taxonomy: A branch of biology that deals with the classification of life on the planet.

Tendon: Connective tissue that attaches muscle to bone.

Test cross: A cross done with a homozygous recessive organism and an organism with an unknown genotype. This cross is done to determine the genotype of the unknown organism.

Testes: The male gonads that produce the sex cells called sperm.

Testosterone: A male sex hormone that is responsible for secondary sex characteristics.

Theory: An explanation to something that has occurred in nature and has been tested many times for validity.

Tissue: A group of cells that perform the same functions in the body of an organism.

Trachea: A tube where air passes from the pharynx to the lungs.

Transcription: The chemical process where complementary copies of sections of nucleotides in the DNA molecules are copied by messenger RNA molecules.

Translation: The chemical process of protein synthesis in the cells.

Transpiration: The process where liquid water enters a plant and is released as water vapor into the atmosphere.

tRNA: (transfer RNA) An RNA molecule that brings the amino acids into the ribosome for protein construction in the process of translation.

Transport: A life process that involves the movement of materials inside the cell, into and out of the cell, and within the body of multicellular organisms.

Triple-beam balance: A device used to manually measure the mass of an object.

GLOSSARY

Trophic level: The name given to one of the various feeding levels present in any food chain within an ecosystem.

Tropism: A directional growth response in plants caused by the release of certain hormones triggered by environmental stimuli.

U

Ultracentrifuge: A device that spins chemicals at a very high speed to separate materials by their densities.

Uracil: A nitrogenous base that is found instead of thymine in the molecule RNA.

Urea: A nitrogen waste product that is present in the urine and sweat of humans.

Ureters: The tubes that connect the kidneys to the bladder.

Urethra: A tube that carries urine from the bladder to outside of the body.

Uterus: The reproductive muscular pouch in the female's body where the embryo develops.

V

Vaccination: The process of injecting a dead or weakened form of a pathogen into the body to initiate an immune response.

Vacuole: Storage sac organelle found inside many cells.

Vagina: The female reproductive structure that connects the uterus to the outside of the female's body.

Valence shell: The outer energy shell of the atom that contains electrons that engage in the chemical bonding process.

Variable: The part of an experiment that is being tested or studied. The variable will be changed during the experiment to show the impact it has on the results.

Vascular bundle: The tubes found in some plants that transports food and water throughout the plant.

Vein: The blood vessel that brings blood under low pressure back to the heart from surrounding body tissue. Some veins contain valves that aid in the movement of blood flow.

Vestige: A structure present in an organism that was functional in some ancestral form but no longer has a function in the present organism.

Villi: The small projections from the lining of the small intestine that increase the surface area for food absorption.

Virus: A very small chemical of DNA or RNA surrounded by protein and capable of reproducing when inside a living cell.

W

White blood cell: The leukocyte cell responsible for the body's defense.

X

X-chromosome: The sex chromosome found in pairs in the female organism and alone in male organisms.

Xylem: A tissue that transports water and minerals in the structure of the vascular plants.

Y

Y-chromosome: The sex chromosome that is only present in male organisms.

Z

Zygote: The diploid cell produced from the fusion of egg and sperm.

INDEX

INDEX

MATTHEW T. DISTEFANO has been teaching high school science for 18 years. Matthew has a BA degree in Biology and a Masters degree in Education from Fordham University. He has taught biology, chemistry, and algebra at Fordham Preparatory School, where he also has served on the technology leadership commitee. Matthew lives in Dobbs Ferry, New York, with his wife, Shawn, and children, Amanda and Luke.